ACROSS THE BOUNDARIES

ENVIRONMENTAL ETHICS AND SCIENCE POLICY SERIES
General Editor: Kristin Shrader-Frechette

IN NATURE'S INTERESTS?
Interests, Animal Rights, and Environmental Ethics
Gary E. Varner

PRIVATIZING PUBLIC LANDS
Scott Lehman

DEMOCRACY, RISK, AND THE COMMUNITY
Technological Hazards and the Evolution of Liberalism
Richard P. Hiskes

ENVIRONMENTAL JUSTICE
Creating Equality, Reclaiming Democracy
Kristin Shrader-Frechette

ACROSS THE BOUNDARIES
Extrapolation in Biology and Social Science
Daniel P. Steel

ACROSS THE BOUNDARIES

Extrapolation in Biology and Social Science

Daniel P. Steel

OXFORD
UNIVERSITY PRESS
2008

OXFORD
UNIVERSITY PRESS

Oxford University Press, Inc., publishes works that further
Oxford University's objective of excellence
in research, scholarship, and education.

Oxford New York
Auckland Cape Town Dar es Salaam Hong Kong Karachi
Kuala Lumpur Madrid Melbourne Mexico City Nairobi
New Delhi Shanghai Taipei Toronto

With offices in
Argentina Austria Brazil Chile Czech Republic France Greece
Guatemala Hungary Italy Japan Poland Portugal Singapore
South Korea Switzerland Thailand Turkey Ukraine Vietnam

Published by Oxford University Press, Inc.
198 Madison Avenue, New York, New York 10016
www.oup.com

Oxford is a registered trademark of Oxford University Press

Steel, Daniel, 1970–
Across the boundaries : extrapolation in biology and social science / Daniel Steel.
p. ; cm. — (Environmental ethics and science policy series)
Includes bibliographical references.
ISBN 978-0-19-533144-8
1. Medicine—Research—Philosophy. 2. Biology—Research—Philosophy.
3. Social sciences—Research—Philosophy. 4. Extrapolation. 5. Causality. I. Title.
II. Series: Environmental ethics and science policy.
[DNLM: 1. Causality. 2. Physiology, Comparative. 3. Population Characteristics.
4. Psychology, Comparative. 5. Public Policy. 6. Research Design. QT 4 S813a 2007]
R852.S72 2007
610.1—dc22 2007007050

9 8 7 6 5 4 3 2 1

Printed in the United States of America
on acid-free paper

For Lynne Anne Steel

Preface

The prehistory of this book began in the late 1990s, when I was a graduate student in the Department of History and Philosophy of Science at the University of Pittsburgh in search of a dissertation topic. I wanted to write something about problems relating to learning about cause and effect in social science, but was having trouble finding a topic that hadn't already been done almost to death and that didn't require mastering impossibly intimidating mathematics. By chance, I stumbled across an edited volume in the Hillman Library titled *Evaluating Welfare and Training Programs* (Manski and Garfinkel 1992), which contained insightful and stimulating discussions of methodological difficulties relating to extrapolating results of pilot studies of welfare-to-work programs. One appealing idea that occurred to me, as well as to some contributors to *Evaluating Welfare and Training Programs,* was that reliable extrapolation often relies upon knowledge of mechanisms linking cause and effect and of factors capable of interfering with those mechanisms. But despite, or perhaps because of, its intuitive obviousness, I had difficulty finding any attempt to elaborate, clarify, or otherwise work out the details of this idea. Here at last was a dissertation topic par excellence! I decided to approach the issue by looking at biological examples that I presumed would best exemplify how mechanisms-based extrapolation could proceed. Figuring out how extrapolation worked in biological cases turned out to be a lot more complicated than I had imagined, and ended up comprising the bulk of the dissertation. As the thoroughly revised, rewritten, and rethought descendant of that dissertation, this book retains the central role of biological examples. Looking back on close to a decade of effort on the topic of extrapolation, I have a sense both of satisfaction in progress made and a painful awareness of the many important questions that remain unanswered. I hope that this book will be a starting point for further progress on methodological problems relating to extrapolation.

Many people deserve thanks for help without which this book would never have come to be. The biggest thanks are due to Sandra Mitchell, who, as my dissertation director, helped me to shape my disorganized ideas into a coherent research project. I am also grateful to Richard Scheines for carefully reading and discussing chapter drafts of that dissertation. There are also a number of people who provided essential assistance in transforming the dissertation into this book. Chief among these is Kristin Schrader-Frechette, whom I met when she came to present a paper at the Michigan State University Philosophy Department. It was largely thanks to her encouragement and support that Oxford University Press agreed to review the manuscript. I'm also grateful to Megan Delehanty for alerting me to the existence of the critiques of animal extrapolation written by Hugh LaFollette and Niall Shanks. Although I disagree with almost all of LaFollette and Shanks's arguments about animal extrapolation, I found them to be the single most useful philosophical source on this topic. It was largely through reading their work that I was able to clearly articulate in simple terms the basic challenges that any adequate account of extrapolation has to surmount. I am also grateful to all those who read and commented on various portions of the manuscript at various stages of completion, including Megan Delehanty, Jason Grossman, Jim Woodward, and Francesco Guala. I would also like to thank the two people who refereed the manuscript for Oxford for their helpful suggestions, and one of those referees especially for suggesting that I change the title. (The original title was *Causality and Heterogeneity*. Try saying that three times fast!) Thanks are also due to audience members for helpful questions asked at presentations at Michigan State University, the University of California at Irvine, the Central Division of the American Philosophical Association, the Center for Philosophy of Science at the University of Pittsburgh, and the Minnesota Center for Philosophy of Science. Finally, I would like to thank Peter Ohlin for prompt and wise editorial advice.

In addition to people, there are some institutions that deserve thanks. I am thankful for the semester of contractual research leave provided by the Department of Philosophy at Michigan State University, and for the Intramural Research Grant Program at Michigan State that enabled me to extend that semester of leave into a full academic year. Without that precious time for reading, thinking, and writing, this manuscript would still be unfinished. I spent that year of research leave as a fellow at the Center for Philosophy of Science at the University of Pittsburgh, which is as about as good an environment for writing a book about philosophy of science as a person could imagine.

Although the bulk of this book is published here for the first time, some chapters contain material from articles that I have previously published. Section 4.4.2 is mostly an abridged version of "Homogeneity, Selection, and the Faithfulness Condition," *Minds and Machines* 16: 303–17. © 2006 by Springer Science+Business Media B.V. All rights reserved. The HIV replication diagram in Chapter 4 (Figure 4.1) and a good deal of Chapter 7 were originally published in "Can a Reductionist Be a Pluralist?" *Biology and Philosophy* 19: 55–73. © 2004 by Kluwer Academic Publishers. All rights reserved. Most of sections 8.2.1 and 8.3.3 originally appeared in "Methodological Individualism, Explanation, and Invariance," *Philosophy of the Social Sciences* 36: 440–63. © 2006 by Sage Publications. All rights reserved. Finally, Chapter 9 is a revised and expanded version of "Social Mechanisms and Causal Inference," *Philosophy of the Social Sciences* 34: 55–78. © 2004 by Sage Publications. All rights reserved.

Contents

ACROSS THE BOUNDARIES

1

Extrapolation and Heterogeneity

Genuine philosophical problems are always rooted in urgent problems outside philosophy, and they die if these roots decay.

—Karl Popper[1]

The best way to introduce the topic of this book is with a few examples.

- Studies find that a particular substance is a carcinogen in rats. We would like to know whether it is also such in humans.[2]
- A randomized controlled experiment has found that a pilot welfare-to-work program improved the economic prospects of welfare recipients. It is desired to know whether the program will be similarly effective in other locations and when implemented on a larger scale.[3]
- On the basis of a controlled experiment concerning outcomes resulting from initiating anti-retroviral therapies earlier or later among HIV+ patients, a physician wishes to decide the best time to initiate this therapy for the patients she treats.

In each of these cases, one begins with some knowledge of a causal relationship in one population, and endeavors to reliably draw a conclusion concerning that relationship in a distinct population. I will use the term *extrapolation* to refer to inferences of this sort. If the populations in question were perfectly homogeneous, extrapolation would be easy: the result from the first population could be directly carried over to the second. But it is not reasonable to assume that the populations in the foregoing examples are homogeneous: they almost certainly differ with respect to characteristics that affect the causal relationship in question. I will use the expression *the problem of extrapolation in heterogeneous populations* (or *the problem of extrapolation* for short) to refer to the challenge of transferring causal generalizations from one context to another when homogeneity cannot be presumed.

The motivation for extrapolation is that evidence concerning the model, or base, population is often more accessible than that for the target with which one is chiefly concerned. For instance, there are many experiments that can be performed on animal models that cannot, for obvious ethical reasons, be performed on humans. However, this only provides a reason why one would want to extrapolate, and does not explain how experimental results concerning a model can be legitimately transferred to a target. Causal relationships in biology and social science typically depend on a variety of conditions that are subject to change, and it is rare that all such factors are known and can be measured. As a consequence, a causal generalization that holds in a given population may be false of a subpopulation or other related populations. The effectiveness of a welfare-to-work program depends on an array of features of the individuals involved, as well as the economic conditions of their locality. Likewise, the effect of an anti-retroviral HIV therapy depends on a range of features of the host as well as of the details of the strain, or strains, responsible for the infection. In both cases, it is unlikely that all of the factors upon which the causal relationship depends can be taken into account in an analysis of the problem. Moreover, the examples provided above illustrate the relevance of extrapolation to such policy issues as regulating the use of a chemical or reforming a social program. This book explores how and under what circumstances reliable extrapolation is possible in biology and social science, and explores some of the implications of this topic for issues in philosophy of biology and social science.

There are several strategies that one might take with regard to extrapolation. The most straightforward is what can be called *simple induction*: infer that a causal relationship found in one population holds approximately in other related populations unless there is some reason to suppose otherwise. Although simple induction is an undeniably important aspect of extrapolation, its limitations are well documented in the toxicology literature (Gold et al. 1992; Hengstler et al. 1999). Simple induction often yields mistaken extrapolations, and it provides no guidance when there is some reason to suspect that the extrapolation might be incorrect. The question is whether there is a more sophisticated account of extrapolation capable of overcoming these limitations.

There are two basic challenges that confront any account of extrapolation that seeks to resolve the shortcomings of simple induction. One challenge, which I call *extrapolator's circle*, arises from the fact that extrapolation is worthwhile only when there are important limitations on what one can learn about the target by studying it directly. The challenge, then, is to explain how the suitability of the model as a basis for extrapolation can be established given only limited, partial information about the target. Critics of animal extrapolation sometimes present this challenge in the form of a vicious circle: establishing the suitability of the model would require already possessing detailed knowledge of the causal relationship in the target, in which case extrapolation would be unnecessary. The second challenge is a direct consequence of the heterogeneity of populations studied in biology and social science. Because of this heterogeneity, it is inevitable that there will be causally relevant differences between the model and the target population. Thus, an adequate account of extrapolation must explain how it can be possible to extrapolate from model to target even when some causally relevant differences are present. Both of these challenges have been posed as general critiques of the methodology of animal extrapolation (cf. LaFollette and Shanks 1993a, 1993b, 1995, 1996).

I argue that earlier work has answered neither of these challenges. There is a small literature that discusses methodological issues relating to extrapolation by reference to detailed case studies (cf. Burian 1993; Ankeny 2001; Schaffner 2001; Guala 2005; Alexandrova 2006). These authors point out that extrapolation is on firmer ground with regard to highly conserved mechanisms, and that the suitability of a model for a particular extrapolation is an empirical hypothesis that must be supported by evidence. But although such methodological observations are undoubtedly correct, they answer neither of the challenges described above. Difficult cases of animal extrapolation typically concern causal relationships—such as the carcinogenic effect of a particular compound—that are not highly conserved. And the observation that the suitability of a model for extrapolation is an empirical hypothesis does not answer the extrapolator's circle. How can that empirical hypothesis be established without already knowing what one wanted to extrapolate? Nor do these studies indicate how extrapolation can be justified when there are some causally relevant differences between model and target.

Others have proposed that *capacities* or *causal powers*—understood as stable influences that are relatively independent of context—can serve as a basis for extrapolation (Cartwright 1989, 1999; Cheng 1997, 2000). The difficulty here is that questions of the stability versus context dependence of a causal relationship are precisely what are at issue in cases of extrapolation. I argue (in Chapter 5) that, when pressed on this matter, existing proposals concerning capacities and causal powers either to revert to simple induction or morph into a version of the mechanisms approach. The mechanisms approach rests on the intuition that knowing how a cause produces its effect provides can provide a basis for extrapolation. It proposes that knowledge of the mechanisms running from cause to effect and of the kinds of things that can interfere with them enhances our ability to reliably decide whether a causal relationship found in one population will or will not obtain in another. This thought is second nature among molecular biologists, and several authors concerned with philosophical questions regarding the role of mechanisms in science have suggested it in passing (cf. Wimsatt 1976, 691; Stinchcombe 1991, 367; Elster 1998, 49; Schelling 1998, 36–37). Yet without further elaboration, the mechanisms proposal does not answer the two challenges described above either. It does not answer the extrapolator's circle, since it is unclear how one can show that the mechanisms in the model are similar enough to the target to justify extrapolation, given the limitations on one's ability to study the mechanisms in the target directly. Moreover, some differences in the mechanism in the model and target are inevitable in biological and social science examples. Thus, the mere invocation of mechanisms does not explain how extrapolation can be justified in the presence of causally relevant disanalogies between model and target.

In this book, I further develop the mechanisms approach to extrapolation so as to more adequately respond to these challenges. I endeavor to clarify the premises that underlie applications of the approach and the types of extrapolative inferences these premises can support, as well as the relevance of extrapolation to some familiar topics in the philosophy of biology and social science. I believe that this project is valuable for several reasons. First, the project is of practical relevance to scientific methodology. A clear understanding of the premises underlying the mechanisms approach to extrapolation helps to reveal the possibilities and limitations of this strategy. If there are circumstances when the requisite premises are problematic, then it is important to know this so as to avoid unreliable applications of the approach. On the other hand, there may be inferences that would be justified by the mechanisms approach that are not being taken advantage of, and an examination of basic assumptions may show in what ways this is the case. Second, I believe that the project is of significant interest for more traditional philosophical topics. As I endeavor to show, a variety of familiar philosophical issues in biology and social science are linked to the problem of extrapolation in heterogeneous populations, including reductionism, ceteris paribus laws, and causality.

The organization of the book is as follows. Chapter 2 presents and explicates a set of concepts—intervention, causal effect, and causal relevance—that recur throughout the remainder of the book. Although my analysis of the first two of these concepts is mostly drawn from other authors, my discussion of causal relevance makes an original contribution insofar as proposing a definition of positive and negative causal relevance that is applicable to cases in which the cause and effect are represented by quantitative variables. It is sometimes claimed that a definition of positive causal relevance should include a criterion of *contextual unanimity*, which requires that a positive causal factor raise the probability of the effect in all background contexts. I argue that this is a mistake, and that such criteria should not be regarded as part of the *meaning* of causal relevance but rather as circumstances that may, when present, facilitate extrapolation.

A mechanisms approach to extrapolation requires an account of the relationship between the qualitative concept of a mechanism and the probabilistic causal notions described in Chapter 2. Chapters 3 and 4 address this issue. The main proposal of Chapter 3 is an account of how mechanisms, on the basis of domain-specific arguments, can be identified with *causal structure*. Often represented by directed graphs, causal structure is that which generates probability distributions and provides information concerning how these distributions will change under interventions. An argument for identifying mechanisms with causal structure in a given context takes the form of an *empirical analysis*, a concept that I draw, with some modifications, from Phil Dowe (2000). A central part of this empirical analysis consists of providing reasons to think that mechanisms are modular in the sense of having independently changeable components. I explain how evolutionary theory can be used to motivate the premise that mechanisms in molecular biology are modular. But although there is a vibrant literature in evolutionary biology on how natural selection may favor modularity, very little has been written on this topic in social science. I make some suggestions about how this evolutionary argument might transfer to social science, but conclude that the case for identifying mechanisms with causal structure is, at present, less well founded in social science than in molecular biology. Further discussion of the circumstances under which social mechanisms can be identified with causal structure is deferred until Chapter 8.

Although the identification of mechanisms and causal structure is an important element of mechanisms-based extrapolation, it is only a first step. Many, and perhaps most, applications of the approach require further, more specific premises about the relationship between probability and causal structure, and hence mechanisms. Chapter 4 articulates a proposition, labeled the *disruption principle*, which plays a fundamental role in mechanisms-based extrapolation of probabilistic causal claims. The disruption principle asserts that interventions on a cause make a difference to the probability of the effect if and only if there is an undisrupted mechanism running from the cause to the effect. After presenting the disruption principle in the abstract, I illustrate it by means of an example drawn from HIV research. Next I consider what justification can be given for the disruption principle. I show that, given the identification of mechanisms with causal structure, it can be derived from two more familiar principles concerning probability and causality, namely, the *principle of the common cause* (PCC) and the *faithfulness condition* (FC). I argue that the aspect of the PCC relevant to the disruption principle rests upon very solid ground but that the case of the FC is more complex. A common objection to the FC is that it is likely to be false when there are counteracting causal paths. I show that such arguments are valid only given a further condition that rarely obtains in heterogeneous populations. Nevertheless, there are some circumstances—such as gene knockout experiments—in which exceptions, or at least *near* exceptions, to the FC are a more serious concern. Hence, this discussion identifies a potential limitation of the disruption principle, and thereby of the mechanisms approach to extrapolation.

Chapters 5 and 6 utilize the concepts articulated in the foregoing chapters to develop an account of mechanisms-based extrapolation. Chapter 5 begins by examining the limitations of extrapolation by simple induction. Next I argue that previously proposed versions of the capacities and mechanisms approaches do not adequately address the two challenges mentioned above: the extrapolator's circle and explaining how extrapolation can be justified in the presence of causally relevant differences between model and target. I then proceed to develop the mechanisms-based answers to these challenges. I begin by explaining how a mechanism in a model organism might serve as a basis for inferring the existence of a corresponding mechanism in the target, by means of what I call *comparative process tracing*. Comparative process tracing relies on background knowledge concerning stages of the mechanism at which significant differences are likely to occur, and where such differences are not likely. In addition, the number of points that must be compared can be reduced further by focusing on downstream stages of the mechanism. Comparative process tracing answers the extrapolator's circle, then, by showing how limited knowledge of the mechanism in the target can suffice to establish the suitability of the model as a basis for extrapolation. I illustrate comparative process tracing by reference to the case of aflatoxin B_1, which concerned the extrapolation of a carcinogenic effect from rodents to humans. Finally, I briefly discuss the relevance of my proposal for disputes concerning the ethical justifiability of animal research.

The second basic challenge confronting an account of extrapolation in heterogeneous populations is that it must explain how extrapolation can be possible even when there are causally relevant differences between model and target. My answer to this challenge is first proposed in Chapter 5, in connection with the aflatoxin B_1 example, and is developed in further detail in Chapter 6, in the context of a discussion of ceteris paribus laws. The central point is that the closeness of match required between model and target depends upon the specificity of the causal claim that one wishes to extrapolate. In particular, a total absence of causally relevant disanalogies is *not* required for extrapolating claims about positive and negative causal relevance. That point is illustrated by the aflatoxin example in Chapter 5, and Chapter 6 articulates some sufficient conditions for extrapolating positive or negative causal relevance. Chapter 6 also discusses a philosophical issue that is closely associated with extrapolation, namely, ceteris paribus laws, which are laws qualified by a clause to the effect of "other things being equal" or "so long as nothing interferes." The expression "ceteris paribus law" is in fact highly ambiguous. Some types of generalizations labeled "ceteris paribus laws" are unproblematic, while the opposite is true for one common interpretation. I explain how the infirmities of the most problematic type of ceteris paribus law vanish if "ceteris paribus" is interpreted as qualifying the extrapolation of positive causal relevance rather than the truth of a universal generalization.

The mechanisms approach to extrapolation is also linked to a perennial issue in philosophy of biology, namely, reductionism. The mechanisms approach to extrapolation operates on the implicit assumption that lower-level details are the place to look for explanations of exceptions to higher-level generalizations. That perspective seems closely tied to reductionism, yet that connection is potentially worrisome, given that reductionism is a highly contentious doctrine. In Chapter 7, I explicate the link between mechanisms-based extrapolation and reductionism. I begin with the suggestion that there are several possible motives for reduction and that different versions of reductionism can be distinguished on the grounds of which goals they aim to achieve. This discussion is used as a basis for clarifying which types of reductionism are presently defended. I then propose that mechanisms-based extrapolation is linked to reductionism just insofar as it implicitly presumes a proposition I call *corrective asymmetry*. Corrective asymmetry obtains when one level of description plays a special role in correcting generalizations at another level, a corrective role which is not reciprocated. I maintain that corrective asymmetry is a criterion of what makes one level more fundamental than another, and hence is a basis for identifying which forms of reductionism genuinely deserve the name. But I also argue that some forms of reductionism that entail corrective asymmetry are compatible with pluralism. In fact, I suggest that corrective asymmetry is helpful for explicating the pluralistic idea that there are autonomous levels of explanation.

Since the best examples of the mechanisms approach to extrapolation I know of come from the biological sciences, the account of extrapolation I propose is developed first in relation to case studies drawn from that domain. Chapters 8 and 9 take up the question of whether the mechanisms approach to extrapolation can be fruitfully extended to social science. There are several challenges confronting this methodological transfer. One is the possibility that social mechanisms do not satisfy the conditions required of causal structure. Chapter 8 picks up the thread of this discussion from Chapter 3. I articulate the concept of structure-altering intervention and explore the circumstances in which interventions are most likely to produce nonmodular changes in social mechanisms. I then turn to a social science example in which extrapolation was a serious concern, namely, the attempt from the mid-1980s to 1990s to estimate the effect of broad-scale changes to the U.S. welfare system on the basis of demonstration programs. Owing to its large scale and relatively unprecedented nature, the intervention in this case was likely to be structure-altering. And in fact, there was a methodological dispute surrounding these studies concerning the value of mechanisms for extrapolating results. I show that a thoroughgoing mechanisms approach, as described in Chapters 5 and 6, is unlikely to be applicable in this case. Nevertheless, I suggest that examinations of social mechanisms in the welfare example are an important supplement to simple induction.

A second challenge for the mechanisms approach to extrapolation in social science is uncertainty about what mechanisms are present. This point is illustrated in Chapter 8 by a case study drawn from experimental economics. The case concerns the extrapolation of a phenomenon known as "preference reversal" from the laboratory to real-world contexts. I show how which of two possible mechanisms is correct has significant implications for how widespread preference reversals are outside the laboratory walls. Chapter 9 examines the challenge of reliably learning social mechanisms. Several authors (cf. Darden and Craver 2001, 2002; George and Bennett 2005) have advanced *process tracing* as a means for discovering mechanisms in biology and social science. Several authors have claimed that process tracing is distinct from and supplements causal inference from statistical data. I argue that existing accounts of how process tracing overcomes challenges confronting causal inference from statistical data in social science are unsuccessful. I then propose a more adequate account that is based on the insight that the appropriate contrast with process tracing is not causal inference from statistical data but rather what I call *direct causal inference.*

2

Interventions, Causal Effects, and Causal Relevance

This chapter presents several concepts—namely, those listed in the chapter title—concerning causality and probability that play a fundamental role in the treatment of extrapolation in heterogeneous populations developed in the remainder of the book. The concept of an intervention has been discussed at length by other authors (cf. Woodward 1999, 2000, 2003; Hausman and Woodward 1999; Spirtes, Glymour, and Scheines 2000), and my presentation of the topic mostly follows these sources. Likewise, I use Judea Pearl's (2000) definition of causal effect, according to which a causal effect of *X* upon *Y* in a population P is a function specifying the conditional probability distribution in P of *Y*, given interventions that set *X* to specific values.

My development of the concepts of positive and negative causal relevance, in contrast, is an original contribution. One important type of extrapolation problem has the following form: We know that *X* is a positive causal factor for *Y* in the population P, and we want to know whether it is also such in the distinct population P′. A systematic inquiry into this inference problem requires a precise and general definition of the expression "positive causal factor." However, such a definition is not to be found in the literature. Philosophical examinations of causal relevance typically treat causality as a relation between events that occur or do not occur, or between properties that are present or absent. Yet many causal relationships of interest to science and ordinary life hold among factors that are naturally represented as varying on a numerical scale: interest rates and rate of inflation; years of education and income; LDL cholesterol level and arterial constriction; fertilizer dosage and plant growth; and so on. Variables representing features of this sort may be called *quantitative*, in contradistinction to those that merely indicate the presence or absence of a property or occurrence or nonoccurrence of an event, which may be called *qualitative*. Christopher Hitchcock (1993, 1995) has shown, though without quite putting it this way, that some traditional philosophical puzzles concerning causal relevance arise from attempting to characterize causal relationships among quantitative variables by means of definitions of causal relevance that are appropriate only for qualitative variables. Hitchcock proposes that claims concerning causal relevance should be understood as providing qualitative information about the causal effect in question (1993, 350).

Although I generally agree with Hitchcock's proposal as far as it goes, it leaves unanswered several questions that need to be resolved before my approach to the problem of extrapolation can proceed. In the case of quantitative variables, just what information concerning the causal effect is provided by expressions that indicate positive (or negative) causal relevance? Moreover, how does an account of positive and negative causal relevance for quantitative variables connect to that for qualitative variables? Presumably, the definition for qualitative variables should be a special case of the one for quantitative variables, but just how is that to work? I undertake to develop an account of causal relevance that answers the above questions. Finally, I consider the suggestion that a requirement known as *contextual unanimity* should be added to any definition of positive and negative causal relevance. I argue that such an amendment would be inappropriate.

2.1 INTERVENTIONS

Interventions are manipulations of something, typically with the intention of bringing about further changes in something else. An intervention might be a complex surgical procedure, the simple act of flipping a switch, or the Federal Reserve's decision to cut interest rates by a quarter of a percent. The concept of an intervention is very useful for drawing the distinction between causation and correlation, a point which can be illustrated by means of an old and familiar example.

We know that there is a statistical association between barometric readings and the occurrence of storms. Let B, A, and S be variables representing barometer readings, atmospheric pressure, and the occurrence of storms, respectively, and let the arrow represent the relationship of direct causation. Of course, the notion of direct causation is relative to the set of variables under consideration, since intermediate causal nodes could be added indefinitely through a continually finer-grained analysis. Then we think that the association between B and S is due to their being effects of the common cause A. Directed graphs consisting of nodes linked by arrows, as in Figure 2.1, will be used throughout to depict *causal structures*. In a directed graph, nodes represent variables (e.g., barometer reading, atmospheric pressure, etc.), while the arrows represent the relationship of direct causation and the absence of an arrow indicates the absence of any causal influence. Thus, the graph in Figure 2.1 says that atmospheric pressure is a direct cause of both barometer readings and storms, but that barometer readings have no influence on storms nor storms upon barometer readings. Causal structures and their relationship to mechanisms will be the topic of discussion in Chapter 3. For the moment, however, a rough characterization of causal structures will have to do. Causal structures refer to complexes of cause-and-effect relationships, as embodied in such things as the electrical wiring in a house, the circulatory system of a human body, or an economy. A graph

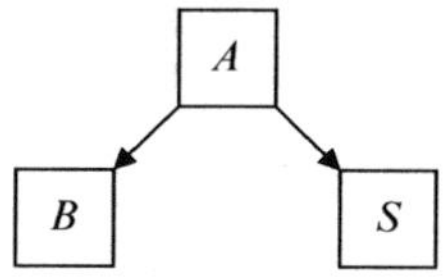

Figure 2.1 Correlation due to a common cause

like that in Figure 2.1 claims to accurately represent some aspect of a causal structure, in this case, one involving a barometer and a meteorological system.[1]

Although the barometer reading is correlated with the occurrence of storms, making it possible to use *B* to imperfectly predict *S*, we do not think that *B* causes *S*. Part of what this judgment means is that the association between *B* and *S* would disappear if we were to intervene as follows. Suppose we place the barometer in a chamber whose air pressure can be set at will, thus allowing us to fix the barometer's reading at any desired level completely independently of *A*. For example, we could randomly choose numbers in the range of possible barometric readings, and then set the reading of the barometer at these values through manipulations of the pressure within the chamber. Under these circumstances, we would expect the probabilistic dependence between *B* and *S* to vanish. On the other hand, if the state of the barometer were (strangely enough) a cause of storms, then we would expect that we could alter the chance of storms by manipulating *B*. This is a commonsense insight regarding causality: interventions on causes yield changes in effects, but not vice versa.

The general concept of an ideal intervention can be abstracted from this simple example. One begins by finding a source of exogenous variation, such as a purely random process such as a coin toss or a roll of dice. The source of variation is exogenous in the sense that, except under the special conditions implemented in the experiment, it is entirely unrelated to the causal process being studied. It comes, as it were, from the "outside." For example, under normal circumstances, barometer readings, atmospheric pressure, and storms are all completely independent of the outcomes of coin flips or rolls of dice. The intervention then consists of arranging the situation so that the source of exogenous variation determines the value of one of the variables in question. For example, given the intervention described in the preceding paragraph, the barometer is no longer affected by the atmospheric pressure, but only by the randomly assigned air pressure inside the vacuum chamber.

An *ideal intervention* can be defined in the following way. Let **V** be a set of variables relevant to a causal structure of interest. Then:

Definition 2.1 (Ideal Intervention):[2] *I* is an ideal intervention on $X \in \mathbf{V}$ if and only if it is a direct cause of *X* that satisfies these three conditions:

(a) I eliminates other influences upon X but otherwise does not alter the causal relations among **V**.
(b) I is a direct cause of no variable in **V** other than X.
(c) I is exogenous.

The intervention is exogenous with respect to **V** just in case it is neither an effect of any variable in **V** nor shares a common cause with any variable in **V**. Intuitively, exogenous causes come from ''outside'' the system. As the barometer example illustrates, randomization is a common way of ensuring that the intervention is exogenous. The intervention in the barometer-storm example can be represented graphically, as in Figure 2.2.

The graph in Figure 2.1 can be called the ''pre-manipulation graph,'' and that in Figure 2.2, the ''post-manipulation graph.'' Note that all three requirements of the definition of an ideal intervention are satisfied in this case. First, the intervention fully determines the value of B, removing all other causal influences, which is represented in this case by the deletion of the arrow from A to B. But aside from obviating any other influences upon the target variable (in this case, B), the intervention leaves all other causal relationships in the original graph unchanged. For example, in Figure 2.2, an arrow from A to S is present, just as in Figure 2.1. Second, the intervention is not a direct cause of any member of the set $\{B, A, S\}$ besides B. Finally, the intervention is exogenous, since it is not an effect of any of these variables nor does it share a common cause with them.

Of course, many actual interventions do not satisfy (a) through (c), and considerable ingenuity and hard work are often needed to ensure that the conditions are approximated in an experiment. Hence, it would be a mistake to suppose that all interventions are ideal. For instance, the Federal Reserve's decision to cut interest rates might be influenced by statistics indicating slowing economic growth, while one of the desired effects of the rate cut is to stimulate the economy. The post-intervention graph representing such a case would contain an arrow running from a variable contained in the pre-manipulation graph to the intervention, in this instance, the rate cut. Thus, the intervention in this example does not fulfill (c) listed above; the decisions of the Federal Reserve are not exogenous to the system that is the target of their interventions. Finally, it should be noted that an intervention need not come about through human activity. An intervention, as defined above, consists in a particular sort of alteration of a causal structure, whether it be brought about by deliberate action or fortuitous circumstance.

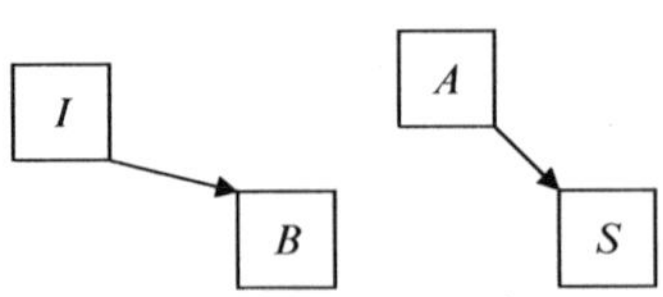

Figure 2.2 An ideal Intervention

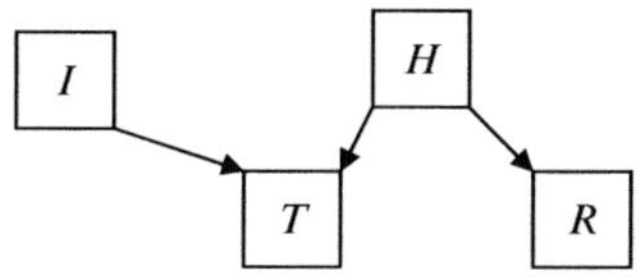

Figure 2.3 An intervention that fails (a)

The concept of an ideal intervention is of interest here primarily in virtue of its usefulness as a basis for defining other causal concepts, such as causal effect and causal relevance. It is very compelling that if X is causally relevant to Y, then ideal interventions on X alter the probability of Y, but otherwise not. For example, it is precisely this assumption that is implicit in randomized controlled experiments, which are generally regarded as the "gold standard" for assessing causal hypotheses. And it is easy to see that standard ways in which randomized controlled experiments can go wrong correspond to a failure of one or more of items (a) through (c).

For example, when some subjects in a clinical trial do not follow the experimental protocol (e.g., do not take the assigned medication as prescribed), then (a) does not obtain. This is problematic, since it allows for the possibility that there is a common cause of the variable being manipulated and the outcome. The sicker patients, for instance, might be less likely to follow the protocol, and also less likely to recover. In such circumstances, there may be a positive correlation between recovery and following the treatment protocol, even if the treatment is entirely ineffective. The point is illustrated in Figure 2.3. In the graph, T indicates treatment; H, health prior to receipt of treatment; and R, recovery.

Likewise, item (b) fails to obtain when the intervention inadvertently directly affects more than one variable in the system. In well-designed clinical trials, for example, great care is taken to ensure that both the test and the control groups are treated identically except that the former receives the treatment and the latter does not. Clearly, item (b) could fail if the intervention provided, along with the treatment, increased confidence in recovery only to those in the test group. Placebos and double-blinds are, of course, standard tactics for avoiding such difficulties. This type of failure of an intervention to be ideal could be represented graphically as in Figure 2.4. In the graph, C is some measure of the subject's confidence of recovery prior to receipt of treatment. In this example, treatment and recovery may be correlated even though the treatment itself is entirely inefficacious.

Item (c) is satisfied whenever I is the product of some purely random process. However, it may fail in the absence of randomization. For example, suppose that the researcher deliberately assigns healthier patients to the test group. Then the intervention is not exogenous, as required by (c). This situation can be represented in Figure 2.5. In such a case,

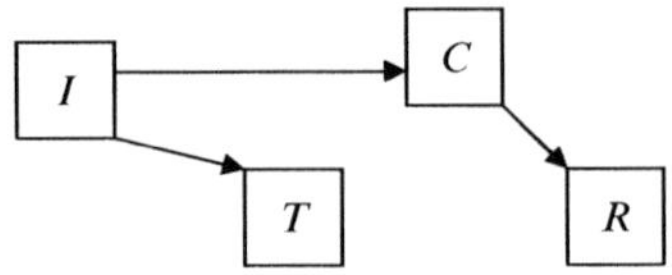

Figure 2.4 An intervention that fails (b)

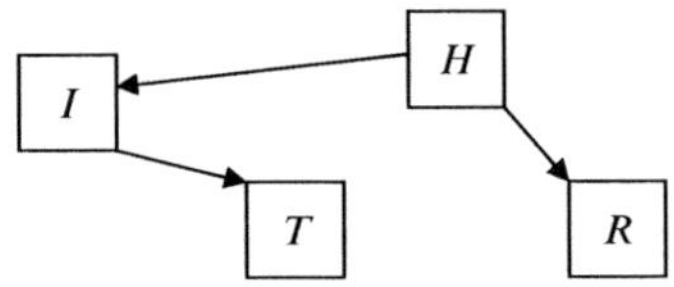

Figure 2.5 An intervention that fails (c)

treatment would be statistically associated with recovery even though the treatment is entirely ineffective.

The guiding intuition of the definitions of causal effect and causal relevance provided in what follows is that one variable, *X*, is causally relevant to another, *Y*, just in case ideal interventions on *X* alter the chance of *Y*. The philosophical aspects of this manipulationist view of causation have been discussed at length in literature on causation (cf. Woodward 2003; Hausman and Woodward 1999). Rather than recapitulate these discussions here, I make only two points. First, I do not claim that the manipulationist view of causation is the only fruitful perspective from which to approach the topic. I choose to rely upon it here because it is a useful and natural manner in which to interpret a wide range of causal claims in biology and social science. In these domains, one often wants to design interventions (e.g., therapies, policies) to achieve desired ends in complex situations in which the outcomes of such interventions cannot be predicted with certainty. Causal effects, defined in terms of the probability distribution of one variable conditional on an ideal intervention on another, are natural objects of inquiry in such circumstances.

Second, the account of causality offered in this section is not intended as a conceptual analysis of causation. A conceptual analysis of causation would consist of a necessarily true statement of the form "*X* causes *Y* if and only if...," where the ellipsis would be replaced with a Boolean combination of terms, all of which can be defined independently of the notion of causality. Since the term "ideal intervention" is itself defined by reference to causation, the account given in this section is not intended as a conceptual analysis.[3]

2.2 CAUSAL EFFECTS

The problem of extrapolation in heterogeneous populations consists in the possibility that a causal effect that holds in a given population may be very different from those holding in subpopulations or distinct, related populations. Thus, the notion of causal effect in a population needs to be clarified before we can get very far with our discussion. The concept of a causal effect is exemplified by controlled in experiments. In such experiments, one is interested in learning the probabilities of certain outcomes (say, recovery and non-recovery) conditional on an ideal intervention that assigns the value of some other variable, say dosage of a drug. Let X and Y be variables representing the treatment and outcome variables, respectively. I will follow the convention of using lowercase letters to denote particular values of variables represented by the same uppercase letters. For example, if X is a variable representing treatment dosage, then x is a particular dosage value, say, 200 milligrams. Pearl (2000) introduces the helpful notation $do(X = x)$, or $do(x)$ for short, to denote an ideal intervention that sets the variable X to the particular value x. Thus, the formula $P(Y \mid do(x))$ is shorthand for a function that specifies the probability distribution of Y conditional on ideal interventions that set X to any particular value x.[4] Given this notation, we can define "causal effect" in the following way (2000, 70):

> *Definition 2.2 (Causal Effect):* For any two distinct variables X and Y, the causal effect of X upon Y is $P(Y \mid do(x))$.

For example, suppose that that X and Y are each binary. Then the causal effect of X upon Y could be represented in a chart, as in Figure 2.6. Here the values of X would be set by an ideal intervention, rather than merely passively observed. Thus, where X represents treatment and Y recovery, this table represents the type of information that is desired from a randomized clinical experiment. It is important to remember that the causal effect $P(Y \mid do(x))$ need not equal the probability distribution of Y conditional on X being passively observed to have the value x, $P(Y \mid X = x)$. These conditional probability distributions may be distinct if Y is a cause of X or there are common causes of X and Y. In those cases, an ideal intervention will eliminate some causal connections between X and Y, which may thereby result in $P(Y \mid do(x))$ being distinct from $P(Y \mid X = x)$. That point is illustrated by the barometer-storm example described above.

Hitchcock (1993, 349) proposes a definition that is very similar to Pearl's, though with a few differences. In Hitchcock's version, the expression defined is not "causal effect" but "the causal relevance of the variable

X	$Y=1$
1	65%
0	27%

Figure 2.6 The causal effect of *X* upon *Y*

X for E," where X is a random variable and E is an event in the technical sense of set theory (i.e., a subset of the outcome space). The causal relevance of the variable X for E is then defined as $P(E \mid X = x)$, where this probability function is assumed to represent the relationship between X and E that holds when all confounding factors have been held fixed. In spite of the similarities, I shall employ Pearl's version. I prefer Pearl's definition because the expression "causal effect" has greater currency than the corresponding phrase defined by Hitchcock[5] and because Pearl's "$do(x)$" notation is very convenient.

For our purposes, one of the most important features of causal effects is that they are prone to vary according to changes in the distribution of factors in the population that affect the outcome. For instance, suppose that we are interested in the causal effect of treatment with penicillin upon recovery from streptococcal infection. This causal effect depends on, among other things, the proportion of individuals in the population who are infected with a resistant strain of the bacteria. In the extreme case in which everyone in the population is thus afflicted, treatment with penicillin may have no effect whatsoever. Since the sensitivity of causal effects to fluctuating features of populations is closely linked to the problem of extrapolation in heterogeneous populations, it is worthwhile to clarify how the term "population" is to be understood here.

Consider the statement that chemotherapy is a positive causal factor for recovery among leukemia sufferers. The most straightforward way to interpret the expression "leukemia sufferers" in this context is to take it to denote a set of real human beings who have suffered, are suffering, or will suffer from leukemia.[6] How far we intend our causal generalization to reach back into the past and extend forth into the future may vary according to several circumstances that will be considered below. The important point, however, is that the generalization is relative to some group of individuals, each of whom exists at some specific time and place in the history of the *actual* world. For example, it would be absurd to object to the proposed generalization by describing a science fiction scenario of a collection of humans whose physiology differed from those of actual people so as to reverse the effect of chemotherapy. But claims about causal effects depend on features of the actual population in other ways as well. Suppose, for instance, that the population of leukemia sufferers consists of two subgroups: one in which chemotherapy causes recovery and one in which it does the opposite. Then the overall effect of chemotherapy will depend crucially on the proportions of these two subgroups and the strength of the effect in each. Indeed, this scenario is hardly far-fetched, given the realworld variability of response to chemotherapy. Thus, I do not assume that populations are homogeneous, that they represent ideal types, or that they constitute natural kinds. It is important for the project of this book that no such assumptions be made about populations, since the problem of extrapolation in heterogeneous populations arises precisely for fields that study populations about which such assumptions are not appropriate.

Statements of causal effect, then, are typically made relative to some actual population of individuals, and the truth of the claim will generally depend on features of that population and the environment in which it is located. Unless specifically indicated otherwise, therefore, the populations of interest for the purposes of this book will be presumed to consist of individuals existing at some place and time in the actual world. The claim that chemotherapy is a positive causal factor for recovery among leukemia sufferers would ordinarily be understood to be relevant not only to people presently suffering from leukemia, but also past and future people so afflicted. However, a researcher might be more or less bold in his or her willingness to extend such a generalization into the future or past. To choose a different medical example, the effectiveness of an antibiotic in the present might not be a reliable guide to its efficacy in the future, since widespread use of it or similar antibiotics would be likely to stimulate the evolution of resistant strains. In such cases, the relevant population would be somewhat vaguely bounded in the future direction. But whether vaguely specified or not, I shall view the populations to which causal generalizations are relative as finite sets of concrete individuals located in specific environments. The individuals in the population need not be contemporaries of each other, but each must exist at some particular time and place in the history of the actual world. Subpopulations, then, are simply subsets of such sets of individuals.

2.3 CAUSAL RELEVANCE

In this section and ensuing subsections, I undertake to develop an account of the concepts of positive and negative causal relevance. Given the definition of causal effect presented above, a definition of causal relevance can be provided as follows.[7]

> *Definition 2.3 (Causal Relevance/Causal Factor):* X is causally relevant to (is a causal factor for) Y if and only if there are values y of Y and x of X such that $P(y \mid do(x)) \neq P(y)$.

This definition makes clear that the bare claim that X is causally relevant to Y is not terribly informative, since it tells us nothing about the manner of this influence. It might be that X promotes or prevents Y, or affects it in some other way. Moreover, the mere statement that X is causally relevant to Y does not tell us which values of X make a difference to Y. For example, X might have little or no effect in low dosages, but a powerful effect in higher ones.

In order to be of practical use, then, ascriptions of causal relevance generally must include some information about the manner in which the cause acts upon the effect. Expressions indicating positive or negative causal relevance are two very common ways to do this. Examples of the first sort of expression are "promotes" and "contributes to," as well as the ordinary usage of "causes." Examples of phrases that can be used to designate negative causal relevance include "prevents," "inhibits," and "blocks." First, I consider the most common proposal for explicating positive and negative causal relevance, namely, the probability-raising definition. After showing why this definition is appropriate only for the case of qualitative variables, I propose a general definition that is applicable in quantitative and qualitative cases alike. Finally, I explain why an additional clause concerning contextual unanimity should not be added to the definition.

2.3.1 The Probability-Raising Definition

Let us begin with the most common probabilistic rendering of such expressions as "A promotes B" or "A contributes to B," namely, the *probability-raising definition* (cf. Suppes 1970; Eells 1991). In such approaches, A and B would typically be interpreted as events or propositions, rather than quantitative variables. Suppose that C represents all common causes of A and B. Then such theories typically assert that A is a positive causal factor for B just in case A is temporally prior to B, and $P(B \mid A \,\&\, C)$ is greater than $P(B \mid \neg A \,\&\, C)$. The probability-raising definition is reasonable in simple examples in which the causes are represented by binary variables, as in classic treatment/non-treatment clinical experiments, but is less adequate when applied to examples involving quantitative variables. The difficulty in question arises in the form of the "problem of disjunctive factors" (cf. Humphreys 1989, 40–41; Eells 1991, 144–68; Hitchcock 1993).

The problem is that when A represents a quantity, the negation of A, $\neg A$, indicates a disjunction of possible values $\{A_1, \ldots, A_n\}$. Moreover, whether $P(B \mid A \,\&\, C)$ is greater than, equal to, or less than $P(B \mid \neg A \,\&\, C)$ can depend on the probabilities $P(A_1), \ldots, P(A_n)$. Consider the following example due to Paul Humphreys (1989, 40–41). We are concerned to test the effectiveness of treatment with a particular drug, A, in bringing about recovery, B. We conduct a randomized controlled experiment in which the subjects are divided into three groups. The first group receives a placebo (A_0); the second, a moderate dose of the drug (A_1); and the third, a large dose (A_2). Suppose that the probabilities in the experiment are the following: $P(B \mid A_0) = .2$, $P(B \mid A_1) = .4$, and $P(B \mid A_2) = .9$. Then, given that $P(A_0) = P(A_1) = P(A_2) = 1/3$, we have $P(B \mid \neg A_1) = .55 > P(B \mid A_1)$. Hence, according the probability-raising definition, moderate doses of the drug prevent recovery. However, this result is quite odd, since the probability of recovery with moderate doses is greater than with a placebo. Moreover, in the example, whether A_1 raises or lowers the probability of B depends on the relative frequency of the treatment assignments. For instance, if $P(A_0) = 7/12$ and $P(A_2) = 1/12$ while all of the other numbers in the example remain the same, then A_1 raises the probability of B.

In general, the problem of disjunctive factors is motivated by the idea that whether X is positively or negatively relevant to Y should not depend upon how frequently particular values of X happen to occur. This intuition is understandable if claims about positive causal relevance are intended to provide information concerning the effects of interventions. If X promotes Y, then increasing X ought to be an effective strategy for increasing Y. But in order for claims concerning positive causal relevance to play this role, it is important that they be invariant under interventions. Furthermore, it is obvious that an intervention normally will alter the probability distribution of the cause, since an intervention seeks to change the distribution of the effect by changing the distribution of the cause. For example, a government health initiative might attempt to reduce the prevalence of lung cancer by reducing the frequency of smoking. Hence, if claims about positive and negative causal relevance are to provide useful guidance concerning the outcomes of interventions, they should be invariant under changes to probability distribution of the cause.

In addition to posing a difficulty for the probability-raising definition of positive causal relevance, the problem of disjunctive factors also provides an argument against using correlation as a measure of positive and negative causal relevance. The correlation between X and Y, $\rho(X, Y)$, is defined as follows:

$$\rho(X,Y) = \frac{\text{cov}(X,Y)}{\sqrt{\text{var}(X)\,\text{var}(Y)}}$$

In this equation, $\text{cov}(X, Y)$ is the covariance of X and Y, which is equal to $E(XY) - E(X)E(Y)$,[8] while $\text{var}(X)$ is the variance of X, which equals $E((X - E(X))^2)$. As long as the values of X and Y consist solely of real numbers, the denominator of the right-hand side of the above equation is equal to or greater than zero. If it is also the case that neither variable is constant (i.e., there is some variation in both X and Y), then the denominator is strictly positive. Hence, for all situations that need concern us here, covariance determines whether the correlation is positive or negative. Yet whether the covariance is positive or negative can depend upon the probability distribution of the cause. For example, consider a case in which X and Y each have three possible values: 0, 1, and 2. Suppose, moreover, that the values of X and Y tend to coincide when $X = 2$, but tend to differ when $X = 1$. In such a case, whether the overall correlation is positive or negative may depend upon the probabilities $P(X = 2)$ and $P(X = 1)$.[9]

One way to deal with problem of disjunctive factors is to propose that attributions of causal relevance are always, though sometimes implicitly, comparisons between probabilities conditional on particular values of the cause (cf. Humphreys 1989; Holland 1986).[10] Hence, in the above example, we could say that moderate doses promote recovery because $P(B|A_0) < P(B|A_1)$. This proposal is quite sensible when the cause is a nonbinary, qualitative variable. For example, suppose one is interested in the influence of race upon employment, where employment is treated as a binary variable and race is treated as a qualitative variable that can take more than two values, say, white, black, Hispanic, or Asian. Suppose that the rate of employment among Asians is highest of all, and that of whites is higher than for both blacks and Hispanics. In this case, whether being white is positively relevant to employment depends on the proportions of the distinct races in the population, in direct analogy to Humphreys's example. It is plausible in this case to insist that claims about positive causal relevance are inherently contrastive, so that the claim that being white is a positive causal factor for employment is understood, implicitly or explicitly, in contrast to being black or Hispanic.

For quantitative variables, however, it is unreasonable to insist that attributions of causal relevance must always be relative to a pair of comparative values of the cause. For example, the claim that smoking causes lung cancer is not equivalent to the statement that, say, that the probability of cancer is greater if you smoke two packs a day rather than just one. The statement that smoking causes lung cancer entails something *in general* about the relationship between cigarette smoking and cancer. A general claim of this sort cannot be identified with a claim about the effects of smoking a specific number of cigarettes any more than the claim that all sparrows have wings can be equated with a claim about a particular bird. Notice that a difference between the smoking example and cases involving qualitative variables (such as the race-employment example) is that it is sensible to speak of the consequences of *increasing* or *decreasing* the cause. The number of cigarettes smoked per day may be raised or lowered; yet it would be nonsensical to speak of increasing or decreasing a person's race.

The problem of disjunctive causal factors suggests that comparisons of the probability of the effect, given specific values of a cause, are at best an adequate account of causal relevance for qualitative variables. That leaves us with the problem of explaining what expressions such as "*X* inhibits *Y*" or "*X* promotes *Y*" mean when *X* or *Y* is a quantitative variable.

2.3.2 Causal Relevance for Quantitative Variables

Stated in terms of the concepts presented above, Hitchcock's (1993) proposal is that claims about positive and negative causal relevance provide qualitative information about causal effects. But what qualitative information, exactly? Consider expressions that indicate positive causal relevance, such as "*X* promotes *Y*" or "*X* causes *Y*." How should such statements be understood when *X* and *Y* are quantitative variables? An appealing idea is that such claims are understood to mean that *increases* in *X* produce *increases* in *Y*. Likewise, "*X* prevents *Y*" means that *increases* in *X* produce *decreases* in *Y*. Hitchcock's discussion of the smoking-cancer example (1995, 261–62) suggests that he, too, shares this intuition.[11] However, this intuitive idea can be interpreted in more than one way. Both interpretations involve reference to some interval of values of the cause. On what I call the *comparative* interpretation, the claim of positive causal relevance indicates that the value of *Y* is greater when *X* is raised from some basal, comparison value to any value within the interval. According to what I term the *monotonic* interpretation, positive causal relevance means that the value of Y increases monotonically with X throughout the interval. The cause might be positively relevant in one of these senses but not the other.

Some preliminary clarification is required to explore these ideas in the present context. In particular, the intuitive idea that positive causal relevance means that increases in the cause yield increases in the effect requires modification for cases in which the relationship between cause and effect is probabilistic. For in that case, increases in the cause do not *always* produce increases in the effect. However, the intuition is naturally extended to probabilistic examples as follows: increases in the cause yield increases in the *expected value* of the effect. Here "expected value" can be understood by means of the notion of an *average causal effect* or, in symbols, $E(Y \mid do(x))$. In the case in which Y is discrete, $E(Y \mid do(x)) = \sum_y yP(y \mid do(x))$. When Y is continuous, $E(Y \mid do(x)) = \int_{-\infty}^{+\infty} yg(y \mid do(x))dx$, where $g(y \mid do(x))$ is a probability density function defined as $P(a \leq Y \leq b \mid do(x)) = \int_a^b g(y \mid do(x))dx$, for any pair of real numbers a and b.

Notice that the average causal effect omits all information concerning the variance of Y, which is an important point, since interventions on X might alter the variance of Y without changing its expected value. For example, imagine a social program that redistributes wealth from rich to poor. Such a program would clearly affect the distribution of wealth in the society but could leave the average, or expected, wealth unchanged. Although the program is causally relevant to wealth, it would be odd to say that the program promotes or inhibits it. However, it would be natural to say that the program promotes economic equality, a thought which is easily accommodated by the present proposal.[12] Thus, I suggest that claims about positive and negative causal relevance are insensitive to changes in the distribution of the effect that leave its mean unaltered. That of course is not to deny that it is important in some circumstances to know how X affects the distribution of Y aside from changing its mean. Rather, the point is merely that such information is not conveyed by claims about positive and negative causal relevance. Note that this situation can arise only if the variable representing the effect is not binary. Given that discussions of causal relevance have tended to focus on binary events or properties, it is not surprising that this complication has not been discussed.

Given these preliminaries, the comparative and monotonic interpretations of positive relevance can be stated more precisely. Let ϑ be an interval of values of X, and let x_0 be some appropriate comparative value of X such that, for every x in ϑ, $x_0 < x$. Then X is a *comparative positive causal factor* for Y within the interval ϑ of X if and only if, for all x in ϑ, $E(Y \mid do(x)) > E(Y \mid do(x_0))$. In contrast, X is a *monotonic positive causal factor* for Y if and only if, for all x in ϑ, $\frac{\partial}{\partial x}E(Y \mid do(x)) > 0$. In other words, X is a monotonic positive causal factor for Y within ϑ when the function $E(Y$

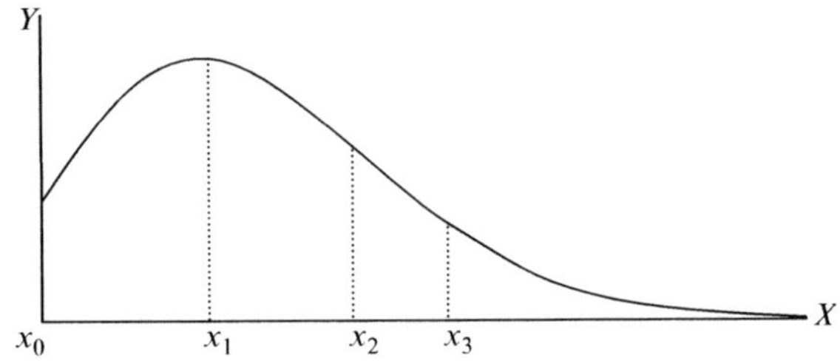

Figure 2.7 The fertilizer example

$| do(x))$ increases throughout ϑ. Definitions of comparative and monotonic *negative* causal relevance can be obtained by substituting "<" for ">" in these two definitions. Likewise, definitions of comparatively and monotonically *neutral* can be obtained by substituting "=" for ">" in the same places.[13]

It will be helpful to illustrate these definitions with a concrete example. Suppose that we are interested in the effect of a certain fertilizer on the growth of a particular species of plant. Let the variable X be a measure of the dosage of the fertilizer and Y a measure of the height of the plant. Imagine that the average causal effect is represented by the curve in Figure 2.7. In the figure, $E(Y \mid do(x))$ increases from x_0 to x_1, where it reaches its maximum; thereafter, $E(Y \mid do(x))$ decreases and ultimately converges to zero.[14] Hence, if x_0 is our comparative value, then X is both a comparative and monotonic positive causal factor for Y within the interval (x_0, x_1). Similarly, X is both a comparative and monotonic negative causal factor for Y in any interval to the right-hand side of x_3. But within the interval (x_0, x_2), X is a comparative positive causal factor for Y, but neither a positive, negative, nor neutral monotonic factor. And within the interval (x_1, x_2), X is a comparative positive causal factor for Y but a monotonic negative factor.

It would be cumbersome and inconvenient, however, to operate with two definitions of positive relevance throughout the remainder of this book. In what follows, I will say that X is a positive causal factor for Y (full stop) exactly if X is both a comparative and a monotonic causal factor for Y.

> *Definition 2.4 (Positive Causal Relevance):* Let ϑ be an interval of values of X, and let x_0 be some appropriate comparative value of X such that, for every x in ϑ, $x_0 < x$. Then X is a *positive causal factor* for Y within the interval ϑ of X if and only if X is both a comparative and a monotonic positive causal factor for Y within the interval ϑ.

The definition of negative causal relevance can be obtained from Definition 2.4 through a simple reversal of inequalities, as explained above. So letting $X = 0$ be our comparative value in the fertilizer example, Definition 2.4 tells us that X is a positive causal factor throughout the open interval (x_0, x_1). Likewise, X is a negative causal factor for Y within the interval x_3 to infinity. Thus, Definition 2.4 coincides with the natural judgment that the fertilizer promotes growth in moderate doses but has the opposite effect in very large doses. Definition 2.4 also nicely treats Humphreys's example that served to illustrate the problem of disjunctive factors. Let Y be a binary variable representing recovery, and let X be a continuous variable representing the treatment dosage. Letting $Y = 1$ stand for recovery and $Y = 0$ for non-recovery, the average causal effect, $E(Y \mid do(x))$, is equal to $P(Y = 1 \mid do(x))$. In Humphreys's example, we know $P(Y = 1 \mid do(x))$ for three values of X, ranging from a zero dosage to a large one. These probabilities suggest that $P(Y = 1 \mid do(x))$ increases monotonically, at least for doses no greater than the largest administered in the experiment. Given that this inference is correct, Definition 2.4 entails that X is a positive causal factor for Y in the interval $(x_0, x_2]$, where x_0 is the zero dosage and x_2 is the large one.

Definition 2.4, then, can be regarded as delineating unambiguous cases of positive causal relevance among quantitative variables. For instance, although the fertilizer is clearly a positive causal factor for growth in moderate doses and a negative factor in very large doses, it is unclear how the effect of intermediate doses should be characterized. For example, when ϑ is the interval (x_1, x_2) in Figure 2.7, $E(Y \mid do(x))$ is greater than $E(Y \mid do(x_0))$ for all x in ϑ, while for all x in ϑ, $\frac{\partial}{\partial x}E(Y \mid do(x))$ is negative. Hence, Definition 2.4 indicates that the fertilizer is neither a positive, a negative, nor a neutral factor for growth in the interval (x_1, x_2), which is a way of indicating the ambiguous nature of the situation. Moreover, given Definition 2.4, any proposition demonstrated concerning conditions in which claims about positive causal relevance can be extrapolated automatically holds for both the comparative and the monotonic senses of that notion. Of course, the practical convenience of Definition 2.4 for the purposes of this book does not show that it represents the one true way to understand positive relevance for quantitative variables. There might be contexts wherein causal relevance is most naturally understood in terms of either comparative or monotonic relevance alone. Nevertheless, I think Definition 2.4 is a reasonable compromise for the present purposes.

An additional nice feature of Definition 2.4 is that it enables us to treat negative and positive causal relevance for qualitative variables as a special case simply by disregarding the monotonicity condition, which is clearly inapplicable in the qualitative case, and by having ϑ be a single value of X rather than an interval. For instance, when X is binary, $x_0 = 0$ and $\vartheta = [1]$ (since $x_0 < x$, for all $x \in \vartheta$). When X is a qualitative variable with more than two possible values (as in the race-employment example), the definition simplifies to the proposal considered in section 2.3.1 that claims about causal relevance are always, explicitly or implicitly, comparisons involving two values of the cause. Thus, the probability-raising definition of causal relevance is a special case of the definition just presented, namely, the case in which X is a qualitative variable. Hence, any proposition that is true of positive, neutral, and negative factors in the case of quantitative variables is also true for qualitative variables (though not vice versa).

Definition 2.4 also captures Hitchcock's (1993, 2003) insight that positive and negative causal relevance are merely two varieties among many. In Definition 2.4, positive, negative, and neutral causal relevance are not collectively exhaustive. For instance, as explained above, in Figure 2.7, X is not neutral with respect to Y in the interval (x_1, x_2), but neither is it a positively nor a negatively relevant causal factor. Note that situations of this kind can arise only if the function $E(Y \mid do(x))$ is nonmonotonic. When $E(Y \mid do(x))$ is constant, monotonically increasing, or decreasing, comparative and monotonic relevance are equivalent.

When $E(Y \mid do(x))$ is a nonmonotonic function, the language of positive and negative causal factors can still be useful (as the fertilizer example illustrates), but may be incapable of describing important aspects of the average causal effect. For example, it is useful to know the value of X for which the function represented in Figure 2.7 reaches its maximum, since this represents the optimum dosage. But this information cannot be expressed in the language of positive and negative causal factors. A similar point can be made with respect to some monotonic cases. For example, suppose that $E(Y \mid do(x))$ increases monotonically and asymptotically converges to the value n. Then it may be important to know the value of n and how quickly $E(Y \mid do(x))$ converges to it, yet such information cannot be expressed in terms of positive and negative causal relevance. In short, the language of positive and negative causal relevance can convey useful information even with regard to nonmonotonic curves, but the more complex the shape of the curve, the more likely that it will be expedient to supplement, or perhaps even replace, talk of positive and negative causal relevance with more detailed descriptions of the shape of $E(Y \mid do(x))$.

The phrase "appropriate comparative value" in Definition 2.4 requires some further comment. In many cases it is very natural to let x_0 be 0; for example, a zero dosage of fertilizer. However, I do not insist that there is one objectively correct choice of the comparative value x_0 in each case. Claims to the effect that X is a positive causal factor for Y serve to provide qualitative information about $E(Y \mid do(x))$. We may wish to convey different sorts of information about the same function in different contexts, and different choices of x_0 may sometimes be useful for this purpose. Notice, however, that in the monotonic case, it makes no difference which value of X we choose for the comparison. For some nonmonotonic functions, the choice of x_0 may be highly arbitrary as well as very relevant to whether X is a positive causal factor for Y, according to Definition 2.4. In such cases, I suggest that the language of positive and negative causal factors is of limited utility.

The only constraint placed on x_0 in Definition 2.4 is that it be strictly less than each point in the interval ϑ.[15] However, there are cases in which it is not implausible that the comparative value x_0 would be greater than every member of the interval. For example, Hitchcock writes:

> We can imagine a country in which almost everyone smokes two packs per day, and in which the surgeon general admonishes citizens to cut back to one pack per day. In such a context, it might be natural to say that smoking (only) one pack per day inhibits lung cancer. . . . (1995, 262)

In this example, the comparative point is two packs of cigarettes per day, while the interval is one pack per day or less. I agree that, in the imagined context, such a choice of interval and comparison point might be convenient for conveying the information that reducing the number of cigarettes smoked from two packs a day to just one reduces the chance of lung cancer. But does this mean that one should say in Hitchcock's example that smoking *prevents* lung cancer? Let us consider what Definition 2.4 has to say about this case.

Observe that Definition 2.4 is not applicable in the case in which the comparative value x_0 is greater than every member of ϑ. Let us consider, then, a modified version of Definition 2.4 in which the comparison point x_0 is greater than every member of the interval. This has the effect of putting the causal claim in terms of the effect that *decreasing* X has upon the expected value of Y. In Hitchcock's example, the envisioned *decrease* in smoking would be expected to produce a corresponding *decrease* in the prevalence of lung cancer. The original version of Definition 2.4, on the other hand, is designed for cases in which causal claims are expressed in terms of the consequences of *increases* in the independent variable. These two modes of expression convey the same information, since decreases in X produce decreases in Y just in case increases in X produce increases in Y. Given that the same information about the function $E(Y \mid do(x))$ is being communicated in both cases, it would be quite misleading indeed to label X's influence upon Y "negative" in one case and "positive" in the other. Thus, if we are to modify Definition 2.4 so that the comparison value may be greater than the values in the interval, we should also reverse the inequality in the definition of comparative causal relevance. Thus modified, Definition 2.4 would state that smoking is a positive causal factor for lung cancer in Hitchcock's example.

With the definition of positive and negative causal relevance, three types of causal claims have been described. In descending order of the precision of the information provided by each type, we have: causal effects, average causal effects, and claims concerning causal relevance. Causal effects and average causal effects can be estimated in some contexts, but are extremely sensitive to changes in background conditions. Consequently, qualitative claims about positive and negative causal relevance are useful in that their roughness and imprecision make them less dependent on the particular circumstances of a specific population. Although it is extremely unlikely that the causal effect found in one heterogeneous population is exactly replicated in another, it may be reasonable to expect that a positive causal factor in one population is also such in other related populations.

2.3.3 Contextual Unanimity

It is sometimes insisted that claims about causal relevance can be properly made only with respect to populations that satisfy a condition called *contextual unanimity* (cf. Cartwright 1983; Eells and Sober 1983; Eells 1986, 1987, 1991). Contextual unanimity obtains when the positive causal factor is such not merely for the population as a whole, but also for every subset of it.[16] However, I shall *not* include contextual unanimity as a part of the definition of positive and negative causal relevance.

Writing contextual unanimity into the definition would make it very hard to see how positive causal relevance could be discovered by the usual scientific means designed for such purposes, particularly randomized controlled experiments (cf. Dupré 1993, 200–1). A randomized controlled experiment may tell us that the cause is positively relevant in the population overall, but such a result is consistent with that effect being neutralized or even reversed in subpopulations. Indeed, among heterogeneous populations it is quite common that there are unknown factors capable of disrupting the mechanism linking cause and effect. Consequently, if contextual unanimity is part of the meaning of claims concerning positive causal relevance, then it is unclear how one could establish that smoking causes cancer, HIV causes AIDS, and so on. In short, if a definition of positive and negative causal relevance is to be applicable to typical examples in biology, medicine, and social science, then it is inevitable that it must allow such claims to be made with respect to heterogeneous populations in which the overall causal effect may be nullified or even reversed in subpopulations. Since claims of positive and negative causal relevance *are* frequently made with respect to heterogeneous populations, it is quite implausible that contextual unanimity is inherent in the meaning of such claims.

Contextual unanimity is best viewed not as a part of the *meaning* of claims concerning positive causal relevance, but as a circumstance that may facilitate extrapolation if present. Adding contextual unanimity to the *definition* of causal relevance is not a fruitful strategy with respect to extrapolation for two reasons. First, as noted above, such an addition would make it practically impossible to learn causal relevance relationships in many areas of biology and social science. Second, although the satisfaction of contextual unanimity can aid extrapolation, it is neither *necessary* nor *sufficient* in general for this purpose.

Extrapolation can be possible even when contextual unanimity does not obtain. In fact, Chapter 6 examines several circumstances that suffice for extrapolating claims about positive causal relevance, *none* of which require contextual unanimity. Consider one very simple example. Imagine a vaccine that is known to be effective in the general population P, although there are some rare cases in which the vaccine has the opposite effect of what is intended. Clearly, contextual unanimity does not obtain in this case. Now consider a proper subset of P, call it P′. We want to know whether the vaccine also inhibits infection in P′. In spite of the failure of contextual unanimity, we would be able to conclude that the vaccine is effective in P′ if we knew that the proportion of negative and positive reactions to the vaccine in P′ is similar to that of the general population P.[17]

Contextual unanimity is also not always *sufficient* for extrapolation. This is most obviously the case when one wishes to extrapolate quantitative information concerning the causal effect, information that may be of practical significance. Even if positive contextual unanimity obtains, for example, the cause may have a strong effect in some populations and a minuscule effect in others. Moreover, there may also be qualitative features of the causal effect that are not expressible in terms of negative and positive causal relevance. For instance, suppose that $E(Y \mid do(x))$ increases monotonically and asymptotically converges to the value n. The value of n, and how quickly the function converges to it, may be important information. Yet even if the population is contextually unanimous, the value of n and the rate of convergence in the population as a whole may differ markedly from that in some subpopulations. This is an extrapolation problem that contextual unanimity, even if it were an available assumption, would not suffice to resolve.

Contextual unanimity is not the only circumstance that might facilitate extrapolation in some circumstances. For example, Chapter 6 examines a condition I call *consonance* that, put roughly, requires that there not be counteracting causal paths from cause to effect. Contextual unanimity and related conditions, such as consonance, should not be viewed as part of the meaning of claims about causal relevance. Instead, they should be regarded as premises that can aid extrapolation in certain types of cases, though not necessarily others. Although I suspect that contextual unanimity is very rarely a justifiable assumption in interesting biological or social science examples, I think that consonance is reasonable in some circumstances. In Chapter 6, I explain consonance in greater detail, consider the circumstances under which it is a reasonable assumption, and examine how it facilitates extrapolating claims concerning positive or negative causal relevance.

2.4 CONCLUSION

Heterogeneity poses a challenge for extrapolation because it raises the possibility that a causal effect in one population might differ in some significant respect from that found in other, related populations. Consequently, clear definitions of "causal effect" and of common expressions for indicating qualitative features of causal effects—particularly, positive and negative causal relevance—need to be given before much progress regarding this problem can be made. This chapter has endeavored to provide these definitions. Let us turn, then, to a consideration of the relation between these probabilistic causal concepts and mechanisms.

3

Causal Structure and Mechanisms

An important prerequisite for exploring the mechanisms approach to extrapolation is to explain what the qualitative concept of a mechanism has to do with probabilistic causal concepts such as causal effect and causal relevance. That is the task undertaken in this chapter and the next. In this chapter, I attempt to show that, for a broad range of cases of interest to the present study, it is reasonable to identify mechanisms with what is called *causal structure* in work on the problem of inferring causal conclusions from statistical data (cf. Glymour and Cooper 1999; Spirtes, Glymour, and Scheines 2000; Pearl 2000; Neopolitan 2004). Accomplishing this necessitates saying something about what causal structure is, and when and why mechanisms can be identified with it.

Explaining how this works involves reconsidering the manner in which analytic philosophers have traditionally approached the topic of causality. One of the primary activities (and perhaps *the* primary activity) of traditional analytic philosophy is conceptual analysis. I understand conceptual analysis to consist of providing necessary and sufficient conditions for the application of an interesting yet somewhat unclear term (e.g., "explanation," "cause"), where these conditions satisfy the following two properties. First, the conditions are stated via concepts that can be defined independently of the target of the definition. Second, the usage of the term recommended by the analysis must agree tolerably well with the intuitions of native speakers in all conceivable circumstances. However, conceptual analysis has decidedly fallen from favor in recent years in the philosophy of science. For example, leading accounts of causality in the recent philosophy of science literature (cf. Hausman 1998; Dowe 2000; Woodward 2003) explicitly disavow any intention to provide a conceptual analysis in the sense just described. Rather than conceptual analysis, these authors endeavor to develop an account of causality that is informed by current scientific theories and methodology. Dowe, whose approach to causation owes much to Wesley Salmon (1984), strives for what he terms an *empirical analysis* of causality, that is, "to discover what causation is in the objective world" (Dowe 2000, 1). Dowe regards current physical theory as the most reliable source of information that would serve as a basis of an answer to this question.

But there is a simple objection to any program that would proceed with empirical analysis before conceptual analysis is complete: without prior conceptual analysis it is unclear what basis there is for asserting that the identified characteristic of the world corresponds to the term derived from ordinary language. David Lewis has posed this objection in the context of a discussion of the philosophy of mind, but it transfers easily to discussions of causation. In Lewis's words:

> Arbiters of fashion proclaim that analysis is out of date. Yet without it, I see no possible way to establish that any feature of the world does or does not deserve a name drawn from our traditional mental vocabulary. (1994, 415)

After considering and rejecting Dowe's response to this objection, I propose that a better answer derives from the view that causal locutions should be treated as theoretical terms in the sense of the Ramsey-Lewis account, according to which theoretical terms are a kind of definite description (cf. Lewis 1970). Given this perspective, an empirical analysis should be based upon a meaning postulate that specifies a particular role associated with the term in question. I will concentrate on two roles ascribed to causal structure; in particular, causal structure is that which generates probability distributions and indicates how these distributions change given interventions.

From this starting point, an empirical analysis of causal structure consists of indicating what fulfills these roles in a particular domain. Making the case for identifying mechanisms with causal structure requires some general argument for supposing that mechanisms are modular, in the sense that it is possible to alter one component without disrupting the functioning of the others. I explain how evolutionary theory can support the claim that modularity is likely to be a pervasive feature of mechanisms. However, this argument is, at present, on firmer ground in molecular biology than in social science, making the motivation for identifying causal structure with mechanisms somewhat more tentative in the latter case. An implication of this discussion is that empirical analyses of causation depend on domain-specific scientific details and hence may differ for distinct phenomena. The question of whether social mechanisms should be identified with causal structure, and under what circumstances, will be explored in further detail in Chapter 8.

3.1 IT'S NICE, BUT IS IT CAUSALITY?

An empirical analysis of causation proceeds by examining the question of what causation is in the world. For example, Dowe's conserved quantity theory advances the following two propositions as the foundation of an answer to that question:

> *CQ1*. A *causal process* is a world line of an object that possesses a conserved quantity.
>
> *CQ2*. A *causal interaction* is an intersection of world lines that involves exchange of a conserved quantity. (2000, 90)

It is striking how removed this analysis is from many ordinary discussions of causation. For instance, it is unclear what relevance exchanges of conserved quantities have to the claim that the vitamin C tablets that Bob ate did not cause him to recover from his cold.[1]

Lewis's objection, then, seems quite apt: the conserved quantity theory is interesting, but why should one regard it as an account of causality? And how can this question be answered without presupposing a conceptual analysis? Dowe responds to this objection in the following way:

> In drawing explicitly on scientific judgments rather than on intuitions about how we use the word, we nevertheless automatically connect to our everyday concept to some extent, because the word cause as scientists use it in those scientific situations must make some historical or genealogical connection to everyday language. (2000, 9)

Thus, basing an analysis of causation on current science connects to commonsense ideas concerning the meaning of "cause" since the usage of the term by scientists is linked to that of ordinary folk. But does this mean that empirical analysis simply amounts to a conceptual analysis of scientists' concept of causation? Dowe makes it clear that this is not his intent: "The task of empirical analysis . . . is not a conceptual analysis of scientists' usage of a term" (2000, 10). Rather, he maintains that the empirical analysis he pursues aims to explicate the concept of causation "implicit in scientific theories" (2000, 11).

The main difficulty I see with this response is that it is highly questionable whether there *is* a concept of causation implicit in current scientific theory. As Dowe observes, no physical theory contains "cause" as an explicitly defined term (2000, 9), and consequently any proposed empirical analysis of causation must inevitably be a substantive thesis over and above what is given by science (Bontly 2006, 182–83). Moreover, there are several ways that one could interpret causation in the light of current science, and it seems unavoidable that arguments for choosing one approach over another will appeal to intuitions about the proper usage of the word "cause." To take just one issue, consider whether causation requires determinism. Dowe argues that the answer is *no*, on the grounds of an example concerning exposure to radioactive material.

> If I bring a bucket of Pb^{210} into the room, and you get radiation sickness, then doubtless I am responsible for your ailment. But in this type of case, I cannot be morally responsible for an action for which I am not causally responsible. (2000, 23)

Thus, given the scientifically plausible assumption that the decay of Pb^{210} is a fundamentally indeterministic process, it follows that indeterministic causation exists.

Although the above argument is interesting and perhaps even persuasive, it is clear that there is more to it than merely explicating a concept implicit in physical theory. Dowe's argument depends crucially on the thesis that moral responsibility (at least in some unspecified class of cases of which the present one is an example) entails causal influence. But what is the basis of any such principle linking moral responsibility and causation? Surely it is not physical theory. Rather, any grounding for it would reside in the interconnection of ordinary concepts of responsibility and causality. As a result, one who maintained that determinism is a fundamental aspect of the concept of causality (e.g., Pearl 2000, 26–27) could avoid the conclusion of Dowe's argument by rejecting the claim that moral responsibility implies causal influence. For example, I might have a moral responsibility to provide assistance to starving people in a distant land despite the fact that I am in no way causally responsible for their unfortunate situation. Thus, Dowe's use of current physics to argue for indeterministic causation requires an antecedent clarification of the relationship between causation and moral responsibility.

Physical theory certainly does have implications for the nature of causation. In the foregoing example, modern physics makes it difficult to maintain both that causation is inherently tied to determinism and that moral responsibility entails causal influence. But this does not show that there is a single account of causality implicit in physical theory, since several different accounts of causation can be made consistent with modern science, depending on what position one takes regarding the interconnections between causation and such things as responsibility, human agency, determinism, temporal priority, spatiotemporal contiguity, and so on. Yet one significant aim of conceptual analysis is to settle questions concerning such interconnections. Hence, we are led straight back to Lewis's objection: empirical analysis cannot fruitfully proceed until matters of conceptual analysis have been settled.

Let us consider a different account of how an empirical analysis of causation can proceed even in the absence of a successfully completed conceptual analysis.

3.2 CAUSALITY AND THEORETICAL TERMS

In this section, I suggest that the cogency of empirical analysis without a successfully completed conceptual analysis can be defended by considering causal locutions as theoretical terms in the sense of the Ramsey-Lewis account (Ramsey 1954; Lewis 1970). The Ramsey-Lewis account proposes to treat theoretical terms as a type of definite description stated via antecedently understood concepts:[2] the theoretical entity is simply that (if anything) which satisfies the description. For example, in eighteenth-century chemistry, phlogiston is that which is present in all flammable objects and is emitted during the process of combustion. In Lavoisier's chemistry, oxygen is that which is absorbed during combustion and is necessary for the formation of acids.

Several authors have suggested that the Ramsey-Lewis account, in addition to applying to deliberately introduced terms of scientific theories, could also be appropriate with regard to concepts falling more squarely in the province of philosophy. For example, Michael Tooley (1987) and Peter Menzies (1996) take such an approach to causation, and Dowe (2000, 49–51) sympathetically considers the idea with respect to transference theories of causation.[3] In Dowe's formulation, such an analysis of causation would consist of three components: a meaning postulate, a contingent hypothesis, and an a posteriori identity (2000, 49). The meaning postulate is the definite description that specifies some important feature of causation: causality is that which does __. For example, one plausible claim is that causation is that which underlies the possibility of predicting the consequences of interventions (cf. Menzies and Price 1993; Woodward 2003). The contingent hypothesis would then be an empirical claim about what things in the world fulfill this role in a given domain, while the a posteriori identification would assert that (in the domain in question) causation is identical to the entity or process indicated in the contingent hypothesis.

The question, then, is how to decide what the meaning postulate should be. An agreed-upon conceptual analysis, if one were available, clearly would be one possible basis for answering this question. For example, Tooley treats his proposal regarding the meaning postulate as a conceptual analysis (cf. Tooley 1987, 25–28). If this were the only possible way to justify one's choice of meaning postulate, then Lewis's argument that empirical analysis cannot proceed until matters of conceptual analysis have been settled would be vindicated. But there is another possibility: the meaning postulate could be derived from empirical observations of the use of causal language. For example, Thomas Bontly proposes that we regard "the concept of causation as a concept defined by its place in an inferential system or network, by the inferences it licenses and those that license it" (2006, 191). Given this perspective, the meaning postulate should be based on inferences that people actually make to and from causation. A meaning postulate, then, should indicate something that is generally regarded as evidence for causal claims as well as something that is judged to be a consequence of causal claims. A meaning postulate that focuses on the connection between causation and predicting the outcomes of interventions does both of these things. The connection between causal claims and effective strategies for achieving ends has been emphasized by many authors (cf. Cartwright 1983, chap. 1; Mellor 1988, 230; Hoover 2001; Woodward 2003). Moreover, carefully controlled interventions are generally regarded as the most reliable scientific means for testing causal claims. There is also experimental evidence that preschool-age children regard interventions as an especially effective way of learning what causes what (Kushnir and Gopnik 2005). Similarly, covariance is generally regarded as a consequence of causal relationships and as evidence for them, at least under the right circumstances (Cheng 1997).

Thus, either manipulation or covariance of the right sort is a potential basis for a meaning postulate in an empirical analysis of causation. In fact, the meaning postulate that will be discussed below—according to which causal structure is that which generates probability distributions and provides information about how they change under interventions—combines both notions. Physical contiguity is a third factor that is often relevant to causal inferences, and it is presumably the guiding thought behind Dowe's conserved quantity theory. However, physical contiguity alone is rarely sufficient to infer causation, since one event might be physically adjacent to another without having caused it. Not surprisingly, in his definition of "C causes E," Dowe combines the definitions of causal process and interaction presented above with a requirement that the cause raise the chance of the effect (2000, 167).

The link between causation and manipulation is doubtful as a conceptual analysis of causation, since specifying what a manipulation or intervention is will inevitably involve references to causation. Nevertheless, a principle linking causation to manipulation can serve as an appropriate meaning postulate for an empirical analysis that treats causation as a theoretical term in the sense of the Ramsey-Lewis theory. If it can be shown that the feature of the world specified in the empirical analysis makes effective manipulation possible, then there is a straightforward answer to the question: Why call it causation? Whatever causation is, knowledge of it is often important for indicating effective and ineffective strategies for achieving ends. Hence, if one identified a general feature of the world that fulfilled this function, then one would have a legitimate claim to be describing causation.

It may be objected that the connection between manipulation and causation could not serve as a meaning postulate, since manipulation is a causal concept, whereas the terms in the meaning postulate are supposed to be antecedently understood. In response, I claim that manipulation and intervention are antecedently understood: they are drawn from the vocabulary of ordinary English and everyday life. (Of course, that does not preclude the usefulness of introducing a framework for discussing them more clearly, as done in section 2.1.) The key point is that *antecedently understood* is a criterion distinct from *independently definable*: we have a reasonably clear idea of what an intervention is, regardless of whether we can define the term in a manner that eschews all reference to causation. Consequently, it is legitimate to use intervention as the basis of the meaning postulate for a Ramsey-Lewis-style definition of "causal structure."

Another possible objection is that without a conceptual analysis of causation, there will be several potential starting points for an empirical analysis of causation. I think it is quite right that there may be several reasonable choices for starting points for an empirical analysis of causation, and that different starting points might lead to separate destinations. However, this is a problem only if one supposes that there must be a monolithic concept of causation for which a unique empirical analysis must be given. In contrast, I see no reason to rule out at the start of inquiry the possibility that the notion of causation is multifaceted.[4] Given the account proposed here, empirical analyses of causation might be pluralistic in two ways. First, a single meaning postulate might be realized differently in distinct domains. For instance, that which generates probability distributions and provides information about how they change under interventions might be one kind of thing in fundamental physics and another in molecular biology and something else again in economics. Second, there may be several reasonable meaning postulates that lead to distinct empirical analyses even within the same domain of inquiry. For example, an empirical analysis based on manipulation might lead to results different from one that emphasizes physical contiguity. The potential for this second type of pluralism raises the question of whether there are common threads linking the several meaning postulates, or whether "causation" is simply an ambiguous term with several distinct meanings. My own view is that the various causal concepts are all closely linked elements of a network of concepts relating to practical reason. However, the account of extrapolation developed in this book does not depend upon the correctness of that overarching vision of causation. All that I require is that the meaning postulate I associate with causal structure be a reasonable one.

Despite the pluralistic spirit expressed in the foregoing paragraph, it is important to stress that not any old thing can be an acceptable meaning postulate. For instance, it would be absurd to say that causation is that which is located in the top drawer of my desk. Absurd proposals like this one would clearly be disqualified by the requirement that a meaning postulate indicate something that is generally regarded as both evidence for and a consequence of causation. But some things that are conceptually linked to causation also fail this criterion. Suppose one proposed this as a meaning postulate: "Causation is that which is necessary for moral responsibility." That there is some conceptual link between moral responsibility and causation seems clear enough. In many cases, one can be morally responsible for something only if one has some influence on it. However, moral responsibility is not something that could serve as *evidence* for causation. Evidence for causation is something that you can actively search for or produce in order to decide whether a causal relationship obtains. If you want to know whether A causes B, you might do an experiment in which you manipulate A and check to see if B varies concomitantly. Or you might collect statistical data to see if A and B are correlated even when potential common causes are statistically controlled for. But there is no analogous way to use moral responsibility as a basis for testing causal claims. The same point would go for the suggestion that causation is that which underlies explanation. Consequently, not everything that is conceptually linked to causation can serve as a good meaning postulate in an empirical analysis of it.

3.3 CAUSAL STRUCTURE

A lively body of work on the problem of causal inference from statistical data uses directed graphs to represent causal structures (cf. Glymour and Cooper 1999; Spirtes, Glymour, and Scheines 2000; Pearl 2000; Neopolitan 2004). For example, consider Figure 3.1.

As in section 2.1, the nodes of the graph correspond to variables and an arrow from one node to another indicates the relationship of direct causation. For instance, *Y* might represent whether or not a particular power strip is switched to the "on" position, while *X* and *Z* each indicate whether or not an electrical appliance plugged into the power strip is on. Using directed graphs to represent causal structures has several advantages for theories of causal inference, the most significant of which is that it enables one to draw upon mathematical results which facilitate computationally tractable methods of deriving predictions about probabilistic independence and conditional independence from alternative causal hypotheses.[5] Directed graphs in conjunction with probability distributions are sometimes referred to as *Bayesian networks*, or *Bayes nets* for short.[6] For convenience, I shall adopt the label *causal Bayes nets* to refer to the approach to causal inference just briefly described.

Causal structures, then, are what directed graphs are intended to represent in the causal Bayes nets literature. But that does not tell us very much about what causal structures are; after all, directed graphs like that in Figure 3.1 can just as easily be used to represent mere correlations. And of course, things other than directed graphs—such as systems of equations and wiring diagrams—can also be used to represent causal structures. What is it, then, that these diverse modes of representation depict? Introductions to treatises on the topic typically emphasize the importance of causal inference for accurately predicting the consequences of public policy decisions (cf. Glymour and Cooper 1999, xi–xii; Pearl 2000, 337; Spirtes, Glymour, and Scheines 2000, xiii–xiv). In addition, significant effort is dedicated to inquiring how knowledge of causal structure, in varying degrees of precision, can serve as the basis of predicting consequences of interventions (cf. Spirtes, Glymour, and Scheines 2000, chap. 7). Thus, causal structures provide information concerning the results of interventions. An additional role is also attributed to causal structures: causal structures are said to "generate" probability distributions (cf. Glymour 1997, 206; Spirtes, Glymour, and Scheines 2000, 29).

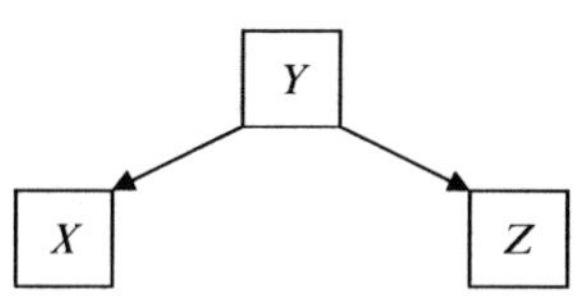

Figure 3.1 A directed graph

Pulling these two strands together, we have the following meaning postulate:

> *(CS) Causal structure* is that which generates probability distributions and indicates how these distributions will change given interventions.

A good understanding of (CS) is evidently dependent on some explication of interventions and of what it is to "generate" a probability distribution. Since the notion of an ideal intervention was explained in section 2.1, let us consider the second of these two questions.

For our purposes, the concern is with physical probability rather than probabilities interpreted as personal degrees of belief or confidence. Although the concept of physical probability is nearly as disputed as that of causation, I think that it is clear enough what sort of phenomena such probabilities usefully represent, namely, processes whose outcomes exhibit what John Venn described as a combination of "individual irregularity with aggregate regularity" (1962, 4). For example, consider the simple case of a flipped coin.

> So long as we confine our observation to a few throws at a time, the series seems to be simply chaotic. But when we consider the result of a long succession we find a marked distinction; a kind of order begins gradually to emerge, and at last assumes a distinct and striking aspect. We find in this case that the heads and tails occur in about equal numbers, that similar repetitions of different faces do also, and so on. In a word, notwithstanding the individual disorder, an aggregate order begins to prevail. (Venn 1962, 5)

As Venn observed, this type of behavior is found in many other circumstances: "Fires, shipwrecks, yields of harvest, births, marriages, suicides; it seems scarcely to matter what feature we single out for observation" (1962, 6).

For our concerns, it is unimportant whether one wishes to define probability as the aggregate or macro pattern itself (as frequency interpretations do), or as the causal tendencies underlying that aggregate pattern (as propensity interpretations do). Probabilities are useful for representing, or modeling, any phenomenon that displays a combination of individual irregularity and aggregate regularity. A process can be said to generate a probability distribution, then, just in case it gives rise to an aggregate pattern of this sort. This criterion is, admittedly, somewhat vague, but it will suffice for the present purposes.

Things that generate probability distributions, then, must exhibit behavior possessing the combination of individual disorder and aggregate regularity described by Venn. I maintain that these properties are possessed by mechanisms that are impinged on by disturbances that are, from the perspective of human knowledge, largely random. Moreover, mechanisms often provide information about the effects of interventions. Consequently, mechanisms are promising candidates for causal structure.

Let us consider this thought in more detail with regard to a pair of cases: molecular biology and social science.

3.4 CAUSAL STRUCTURE IN MOLECULAR BIOLOGY

Given a meaning postulate, the next stage of an empirical analysis is a contingent hypothesis, which specifies a class of entities whose extension, in a particular domain, is exactly that of the meaning postulate. In this section, I argue that in molecular biology, causal structure coincides with mechanisms, yielding the following empirical analysis:

- *Meaning Postulate* (CS): *Causal structure* is that which generates probability distributions and indicates how these distributions change under interventions.
- *Contingent Hypothesis*: In molecular biology, *mechanisms* are what generate probability distributions and indicate how these distributions change under interventions.
- *A Posteriori Identity*: In molecular biology, mechanisms *are* causal structure.

In this section, I argue in favor of the above contingent hypothesis. As explained in earlier sections of the chapter, empirical analyses rely upon established scientific theories of the relevant domain. In this case, evolutionary biology plays an important role in motivating the claim that mechanisms in molecular biology provide information about the consequences of interventions by providing a general reason to expect that such mechanisms are modular.

3.4.1 What's a Mechanism?

Mechanisms, in a very literal sense of the term, are paradigmatic examples of causal structures. For example, in Nancy Cartwright's words:

> The car engine is a good case of a stable causal structure that can be expected to give rise to a probability distribution over the events of the cooperating causal processes that make it up. That is why it can make sense to ask about the conditional expectation of the acceleration given a certain level of the throttle. (1995a, 72)

Given that several authors have proposed that mechanisms play an important role in the life sciences (cf. Bechtel and Richardson 1993; Glennan 1996; Machamer, Darden, and Craver 2000), they are a natural place to turn for an empirical analysis of causal structure in biology. However, this must be done with some care, since the application of the word "mechanism" in distinct domains might reflect only a superficial similarity of subject matter. Thus, it is important to examine just what sorts of things biological mechanisms are and why they should be thought to fulfill the roles ascribed to causal structure.

Mechanisms are generally understood as consisting of interacting components that generate a causal regularity between some specified beginning and end points. For example, according to a definition proposed by Peter Machamer, Lindley Darden, and Carl Craver, "Mechanisms are entities and activities organized such that they are productive of regular changes from start or set-up to finish or termination conditions" (2000, 3). This general characterization is appropriate for literal examples of mechanisms, such as the car engine, and is reasonable with regard to things referred to by the term "mechanism" in biological science. Consider, for example, the mechanism involved in protein synthesis, in which the series of nucleotide bases in strands of DNA influences the chemical structure of proteins produced within cells. Nearly any introductory biology textbook describes this mechanism roughly as follows. First, a strand of DNA unwinds and the adjoining nucleotide bases separate. The next step is the transcription of the unwound DNA by messenger RNA (mRNA), the order of the bases of the mRNA being determined by the order of the complementary nucleotide bases in the DNA strand. Finally, the strand of mRNA serves as a template for transfer RNA (tRNA), which assembles a string of amino acids into a protein. In this case, the interworking parts give rise to more readily observed regularities, such as correlations between genes and specific traits. Some things referred to by the term "mechanism" may not involve a regular series of changes. For example, the term "mechanism" is sometimes used to refer to a unique chain of events leading to a particular effect. However, since this book is concerned with extrapolating causal generalizations, I will use the term "mechanism" to refer to regularly operating causal relationships rather than idiosyncratic and unique chains of events. Consequently, I will restrict the term "mechanism" to processes that satisfy the "regular changes" clause of the Machamer-Darden-Craver definition.

Other related definitions of mechanisms exist. For example, Stuart Glennan proposes a definition that is similar to Machamer, Darden, and Craver's except that it requires that the interactions among the components of the mechanism be governed by "direct causal laws" (1996, 52). The reference to laws in this definition is problematic, since it is debatable whether there are genuine laws of nature in biology and social science, where the term "mechanism" is often used. Consequently, in a subsequent revised account of mechanisms, Glennan replaces "direct causal laws" with "direct, invariant, change relating generalizations" (2002, S344). The notion of an invariant generalization is borrowed from James Woodward (2000, 2003). An invariant generalization is one that is invariant under some range of ideal interventions on the allegedly explanatory variable. For example, the generalization that barometer readings and storms are correlated is not invariant under ideal interventions on the barometer readings (as explained in section 2.1). Hence, the barometer readings do not cause or explain storms, according to Woodward's theory. In contrast, the generalization that smoking is correlated with lung cancer would be invariant under ideal interventions that target smoking. James Tabery (2004) argues that there is an important difference between Woodward's conception of causation and the notion of "productivity" invoked in the definition proposed by Machamer, Darden, and Craver. The thought is that while invariant generalizations merely point to ways in which changes brought about by an intervention lead to specific changes someplace else, productivity pertains as well to cases in which new entities are constructed (2004, 8–9). However, the "changes" covered by Woodward's account of causation should be understood to include constructing a new product out of disparate parts. For example, imagine a cellular process that generates a particular enzyme. Let *E* be a variable that indicates whether or not this enzyme has or has not been produced on given occasions. Then there may be invariant generalizations relating *E* to other variables that represent, say, the presence of necessary components in the cell or the transcription of a particular gene. If there is a real difference between Glennan's definition and that proposed by Machamer, Darden, and Craver, I think it is only that Glennan provides more detail about his preferred interpretation of causation.

Cartwright's *nomological machine* is another mechanism concept. Cartwright defines a nomological machine as "a fixed (enough) arrangement of components, or factors, with stable (enough) capacities that in the right sort of stable (enough) environment will, with repeated operation, give rise to the kind of regular behaviour that we represent in our scientific laws" (1999, 50). Like the definitions of mechanism considered above, Cartwright's nomological machine consists of interacting components that generate causal regularities. The concept of a nomological machine is distinctive only insofar as it is founded on Cartwright's concept of a capacity. A capacity is a stable causal power that exerts its characteristic influence in a broad range of contexts (Cartwright 1989, Chapter 4). The pure effects of a capacity can be observed only in special experimental circumstances in which all other causes have been eliminated, but the capacity nevertheless makes its contribution to the effect even when other causes are present. Since Cartwright regards physical laws merely as descriptions of the behavior of a capacity in the idealized situation in which no other forces are acting, she regards capacities as ontologically more basic or fundamental than laws of nature. Cartwright also argues that interpreting causal relationships by reference to capacities is essential for understanding how it is possible to extrapolate causal claims from one context to another (1989, 163). I argue in Chapter 5 that capacities do not in fact have this special virtue. But for the moment, let us sum up the above survey of mechanism concepts.

All of the definitions canvassed above characterize mechanisms as consisting of sets of interacting components that generate a regular series of causal interactions. To the extent that they disagree, it is with regard to how to interpret causation. For example, Glennan's original definition (1996) characterized causation by reference to "direct causal laws," while

Cartwright prefers capacities. Fortunately, pursing an empirical analysis of causal structure does not require deciding whether laws or causal powers are more fundamental or insisting that there is one correct way to interpret causation. Instead, it requires an argument that mechanisms generate probability distributions and provide information about how those distributions change under interventions. Given this, I will adopt the Machamer–Darden–Craver definition, since it is the least specific about causation, laws, and their relation to one another. The question, then, is whether mechanisms, so defined, are causal structures. I consider this question first with regard to molecular biology and then for social science.

3.4.2 Mechanisms, Modularity, and Evolvability

There is good reason to think that if there is such a thing as causal structure in molecular biology, it would have to be mechanisms. First, note what might be called the working assumption of molecular biology: all causal relationships in living organisms are mediated by molecular processes. This working assumption rests on the attractiveness of physicalism as a general ontological principle and on the success of molecular biology as a research program. Thus, if mechanisms are not causal structures in molecular biology, it is hard to see what could be. However, this conclusion is only half of the argument. It is also necessary to show that mechanisms in molecular biology do in fact perform the functions required of causal structure.

Since causal structure is that which generates probability distributions and provides information about how those distributions change given interventions, there are two parts to this argument. Let us begin with the requirement that causal structure generate probability distributions. Is this something that mechanisms in molecular biology do? Recall the features that Venn judged to be characteristic of phenomena to which the concept of probability can be usefully applied: individual disorder combined with aggregate regularity. It is obvious that mechanisms in the sense of the Machamer–Darden–Craver definition will tend to generate large sample regularities, given the requirement that mechanisms "are productive of regular changes" from the beginning and end stages of the process. Moreover, biological mechanisms are invariably subject to an array of disturbing influences, many of which are not well understood. Thus, from the perspective of human knowledge, individual cases of the operation of a given mechanism in molecular biology will inevitably display a certain amount of random variation, which is an example of the "individual disorder" that Venn described. Notice that the same sort of situation is found in the case of human-constructed machines, which are often given as paradigm examples of causal structure. They produce regular changes, yet are impinged upon by a variety of disturbing influences that often cannot be known with any exactitude. Consequently, we have a straightforward account of why mechanisms in molecular biology should display the aggregate regularity and individual disorder that Venn cited as the characteristic features of probabilistic phenomena. Of course, these aggregate patterns may themselves change in the course of evolution, but this simply illustrates the familiar point that probability distributions themselves can change over time (cf. Venn 1962, 14–17). This point is illustrated by such social statistics as the marriage rate or average life span. Indeed, it is exemplified by Cartwright's case of the car engine; the probability of a breakdown increases as the engine ages.

However, since knowledge of causal structure also provides information about the consequences of interventions, an account of why mechanisms in molecular biology should be thought to generate probability distributions is only half of the story. It is necessary to argue that mechanisms in molecular biology generally provide information about the results of interventions. On the face of it, it is quite plausible that this is the case. Indeed, this presumption that knowledge of mechanisms can indicate the consequences of various types of interventions is often the reason for trying to discover them. But is there some general feature of biological mechanisms that justifies this presupposition? One answer to this question has been suggested by Woodward (2002a, S374–76), who maintains that mechanisms are *modular* in the sense that it is possible to intervene to change a feature of one component while leaving the generalizations that govern the others unaltered. This idea is reflected in the manner in which interventions are represented in directed graphs. Consider again the case of the two appliances plugged into the same power strip, represented by the graph in Figure 3.1. Recall that an ideal intervention takes complete control of the variable it targets (say, *X*), so as to eliminate all other influences that otherwise affect it. Such an intervention, as we saw in section 2.1, would be represented as shown in Figure 3.2.

Of course, many real-life interventions are not ideal. In our example, switching on one of the appliances would not be an ideal intervention, since it does not sever the influence of the state of the power strip. Such an intervention might be represented as shown in Figure 3.3:

The important point with regard to modularity in figures 3.2 and 3.3 is that besides possibly eliminating or weakening the influence of *Y* upon *X*, the intervention leaves all other causal relationships unaltered. For example, modularity would be violated if the intervention eliminated the influence of *Y* upon *Z* or created a causal chain from *X* to *Z*. The interest in modularity stems from the fact that it facilitates predicting the

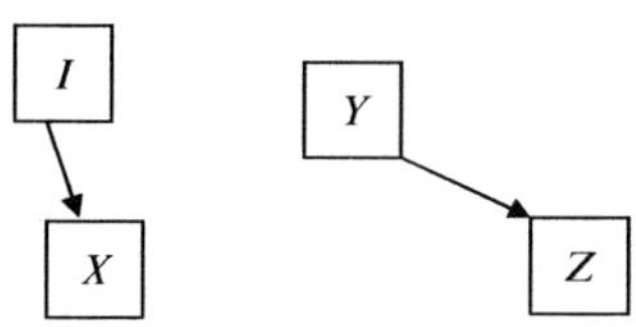

Figure 3.2 An ideal intervention

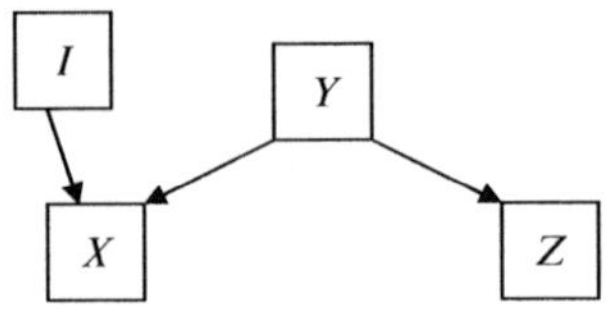

Figure 3.3 A nonideal intervention

consequences of interventions, since except for the elimination of influences upon the targeted variable, the causal structure operates as before. Thus, the persisting causal structure can be used to trace out the intervention's consequences. Given Woodward's proposal, the question is whether there is reason to believe that biological mechanisms are usually modular. Some classic examples of biological mechanisms do exhibit modularity. For example, it is possible to alter the sequence of nucleotide bases in a functional segment of DNA while the other components in the mechanism of protein synthesis continue to function as before. But is modularity only an adventitious feature of some restricted class of biological mechanisms, or is there some reason for supposing that it obtains in general?

One way to argue that modularity is likely to be a commonly occurring feature of biological mechanisms is to maintain that modularity is favored by natural selection. Herbert Simon (1962) was one of the first to suggest a general explanation of how modularity is adaptively beneficial in environments in which disruption or interference is common. The proposal can be illustrated with a modified version of one of Simon's best-known examples, the parable of the expert watchmakers Hora and Tempus (1962, 470).[7] Hora constructs her watches by building independently changeable modules that can be assembled into the final product. In contrast, Tempus constructs holistic watches in which no part can be modified independently of any of the others. Hora's modular production method gives her an advantage over Tempus as their ever more popular watches are used in new circumstances. For instance, mountain climbers find that the watches of Hora and Tempus fail to operate properly at high altitudes. Hora is able to trace the problem to a specific module, and through trial and error she develops a new version that operates properly under high-altitude conditions. In contrast, Tempus must redesign an entirely new high-altitude watch, which means searching for a solution through the space of possible watches, which is far vaster than the space of possible modifications of a specific component. By the time Tempus has finished his holistic high-altitude chronometer, Hora has already cornered the mountain climber watch market, as well as that for scuba divers, mariners, pilots, runners, and several other specialty niches. The moral of the parable, then, is that modularity facilitates finding quick solutions to new problems, which is essential for adapting to changing environments.

The theme of this parable is nicely illustrated by the HIV replication mechanism that will be discussed in detail in the next chapter. HIV is notorious for its ability to evolve resistance to drugs designed to block its replication. Typically, such drugs interfere with one stage of the replication mechanism, for example, by binding to and disabling an enzyme required for a step in the process. In this case, a mutation in the viral genome can result in a slightly modified version of the enzyme to which the therapeutic compound no longer binds. Given that the other components of the mechanism continue to function as before, HIV has successfully evolved resistance; but if the change to the enzyme resulted in cascading alterations to the other components, it is likely that the mutant strain would no longer be viable. Thus, the HIV replication mechanism is analogous to Hora's production method: since it is modular, alterations to one component to do not compromise the functionality of the others. Consequently, evolving resistance to a single drug requires altering only one component of the replication mechanism, and hence searching through a smaller space of possibilities. In contrast, if the HIV replication mechanism were holistic like Tempus's watches, evolving resistance to the therapeutic compound would require rebuilding the mechanism from scratch, and hence searching for a solution in the space of all possible HIV replication mechanisms. Thus, modularity is an important part of what enables HIV to quickly evolve resistance.

These examples suggest that modularity enhances fitness by promoting adaptability to changing environments. Moreover, environmental perturbations of various kinds—new predators, changes in supply of resources, and so on—are a pervasive fact of life. Hence, evolutionary theory suggests a basis for expecting that modularity is a typical characteristic of biological mechanisms. In fact, the importance of modularity to adaptability is a familiar point in evolutionary biology (cf. Wagner and Altenberg 1996). There is a growing body of theoretical work that attempts to clarify the general mechanisms whereby natural selection could give rise to modularity (cf. Ancel and Fontana 2000; Lipson et al. 2002; Kvasnicka and Pospichal 2002; Kashtan and Alon 2005). This work supports the intuition that natural selection favors modularity in changing environments, but with some refinements. For example, one recent study suggests that although not all varying environments lead to modularity, modularity is favored in environments with "modularly varying goals" (Kashtan and Alon 2005, 13777). Goals vary modularly when new goals share subproblems with preceding goals (ibid., 13775). The HIV example illustrates this concept. At first, the goal of the enzyme is to achieve a particular function, say, to reverse transcribe viral RNA to DNA. After the start of the drug treatment, the enzyme must still perform its original function while also avoiding being bound to the therapeutic compound. Hence, reverse transcribing the viral RNA to DNA is a subproblem shared by the first and second goals. The situation in the watchmaker parable is similar. In redesigning the malfunctioning module, Hora must preserve its original function while avoiding the disruption that occurs at high altitudes. Modularly varying goals might drop as well as add subproblems. For instance, consider a population of fish that has colonized a network of underground pools: the fish no longer need to see, but they still need to swim.

There is also a growing number of empirical studies that examine the role of modularity in the evolution of particular lineages (cf. Beldade et al. 2002; Chipman 2002; Mabee et al. 2002; Friedman and Williams 2003; Emlen et al. 2005; Fraser 2005).[8] These studies provide fascinating concrete examples of the ways in which modularity can be manifested in living beings. For example, one study documents how threshold mechanisms allow for developmental modularity in the evolution of beetle horns (Emlen et al. 2005). Empirical studies can also test hypotheses about the relationship between modularity and evolvability. For instance, mixing and matching modules, sometimes called "compositional evolution," may often be a more efficient means of finding a solution to a problem than randomly rearranging basic elements (Watson and Pollack 2005, 456). By analogy, one is more likely to produce a sentence by randomly combining clauses and phrases than by randomly combining letters and spaces. An additional potential advantage of compositional evolution, in contrast to gradual accumulation of slight variations, is that it can avoid suboptimal local maxima traps, since a rearrangement of modules constitutes a jump to a nonadjacent point in the fitness landscape (Kashtan and Alon 2005, 13777). And in fact a recent study finds support for compositional evolution with regard to protein modules in yeast (Fraser 2005). In the HIV example discussed above, compositional evolution would suggest that the resistant variant resulted from rearranging proteins that compose the enzyme rather than from shuffling the individual amino acids that make up the proteins.

Mechanisms that are modular in the sense of these biological discussions are ipso facto a useful basis for predicting the consequences of interventions. Although several modularity concepts can be found in biology (Schlosser and Wagner 2004), the following is a fairly standard, rough definition that is appropriate for the present context:

> A modular representation of two character complexes C1 and C2 is given if pleiotropic effects of the genes fall mainly among members of the same character complex, and are less frequent between members of different complexes. (Wagner and Altenberg 1996, 971)

According to this definition, modularity states that the multiple effects of genes tend to focus on discrete trait complexes. This definition makes the connection between modularity and manipulability straightforward. For if modularity in the sense just defined obtains, it is possible, by means of appropriate alternations to the genome, to intervene to alter one component of the mechanism without significantly disturbing the others. Thus, knowledge of modular mechanisms would provide information about the consequences of interventions. Of course, it would be a mistake to take the above as a *general definition* of modularity. Rather, it is a rough specification of the physical basis of modularity in molecular biology—in effect, an empirical analysis of modularity in that context. An empirical analysis of modularity in social science, for instance, would have to be something rather different.

In sum, given the meaning postulate that causal structure is that which generates probability distributions and indicates how such distributions change given interventions, evolutionary theory plays a central role in an empirical analysis of causal structure in molecular biology. Evolutionary theory can be invoked to support the claim that in the context of molecular biology, mechanisms can be identified with causal structure, since it provides an account of why it should be expected that biological mechanisms are typically modular. Modularity, meanwhile, was linked to the ability to predict the consequences of interventions. Of course, since empirical analysis depends on current scientific theory, it is inherently tentative. New scientific developments might result in significant revisions to the theory, and these developments might have implications for the empirical analysis. The evolution of modularity in biological systems is a young and thriving research area, which means that we should expect surprises yet to come.

3.5 CAUSAL STRUCTURE IN SOCIAL SCIENCE

In this section, I consider the possibility that an empirical analysis identifying causal structure with mechanisms in molecular biology on the basis of evolutionary theory could work similarly in social science. On its face, the argument for the adaptive benefits of modularity in variable environments seems entirely general, and hence applicable to cultural as well as to biological evolution. However, the details of these proposals are at present far less developed in social science than in biological science. In addition, one common argument against the possibility of laws of social science can be interpreted as an attempt to show that social mechanisms will often respond in nonmodular ways to interventions. Thus, I conclude that although it is likely that the evolutionary account of modularity described above can be applied to some social mechanisms, the extent to which this is so is even more of an open question than in the case of mechanisms in molecular biology.

3.5.1 What's a Social Mechanism?

In order to consider whether social mechanisms are likely to be modular, some clarification of "social mechanism" is called for. Earlier, mechanisms in general were roughly characterized as sets of entities and activities organized so as to produce a regular series of changes from a beginning state to an ending one. Social mechanisms in particular are usually thought of as complexes of interactions among agents that underlie and account for macrosocial regularities (cf. Little 1991, 13; Stinchcombe 1991, 367; Schelling 1998, 33; Gambetta 1998, 102). The paradigm example of an agent is an individual person, but coordinated groups of individuals motivated by common objectives—such as a corporation, a government bureau, or a charitable organization—may also be treated as agents for certain purposes (cf. Mayntz 2004, 248). Social mechanisms are sometimes tied to the assumption that the agents comprising them are rational, say in the sense of being utility maximizers. For instance, Tyler Cowen writes, "I interpret social mechanisms . . . as rational-choice accounts of how a specified combination of preferences and constraints can give rise to more complex social outcomes" (1998, 125). I shall not adopt this perspective, and hypotheses about social mechanisms will not be restricted to rational-choice models.

Social mechanisms typically involve reference to some categorization of agents into relevantly similar groups defined by a salient position their members occupy vis-à-vis others in the society (cf. Hernes 1998; Little 1998, 17; Mayntz 2004, 250–52). In the description of the mechanism, the relevant behavior of an agent is often assumed to be a function of the group into which he or she is classified. For example, consider the anthropologist Bronislaw Malinowski's (1935) account of how having more wives was a cause of increased wealth among Trobriand chiefs. Among the Trobrianders, men were required to make substantial annual contributions of yams to the households of their married sisters. Hence, the more wives a man had, the more yams he would receive. Yams were the primary form of wealth in Trobriand society, and served to finance such chiefly endeavors as canoe building and warfare. Although individuals play a prominent role in this account, they do so as representatives of social categories: brothers-in-law, wives, and chiefs. The categorization of component entities into functionally defined types is not unique to social mechanisms. Biological mechanisms (e.g., that of HIV replication) are often described using such terms as "enzyme" and "co-receptor." The terms "enzyme" and "co-receptor" resemble "chief" and "brother-in-law" in virtue of being functional: all of these terms provide some information about what role the designated thing plays in the larger system of which it is a part. In sum, social mechanisms can be characterized as follows. Social mechanisms are complexes of interacting agents—usually classified into specific social categories—that produce regularities among macrolevel variables.

This characterization of a social mechanism can be illustrated by another, better-known example. Consider Thomas Schelling's bounded-neighborhood model, which is intended to account for persistent patterns of segregated housing in spite of increased racial tolerance (Schelling 1978, 155–66). In this model, the residents of a given neighborhood are divided into two mutually exclusive groups (e.g., black and white). Each individual prefers to remain in the neighborhood, provided that the proportion of his or her own group does not drop below a given threshold, which may vary from person to person. Meanwhile, there is a set of individuals outside the neighborhood who may choose to move in if the proportions are to their liking. Clearly, this model divides individuals into groups with which characteristic preferences and subsequent behavioral patterns are associated, and by these means accounts for macroregularities.

On the face of it, it might seem that the empirical analysis of causal structure given in section 3.4 easily transfers to social science. As in the case of molecular biology, it is difficult to see what could constitute causal structure in social science if not social mechanisms. Moreover, it is plausible that social mechanisms often produce stable patterns, and hence generate probability distributions. Finally, just as in the case of biology, it seems that modularity is a feature that contributes to the adaptability of social systems. Indeed, the parable of Hora and Tempus illustrates the advantages of modularity for technology and is analogous to such historical cases as the IBM PC versus the Apple Macintosh, and General Motors versus Henry Ford (cf. Langlois 2002, 23–33). However, it is unclear how far the evolutionary argument for the prevalence of modularity carries over to the social realm.

3.5.2 Modularity and Social Mechanisms

Let us consider how the evolutionary argument for modularity described in section 3.4.2 might work with regard to social phenomena. As a first stab, consider the following suggestion. Modular social mechanisms contribute to the adaptability of the social groups containing them. Such groups would be able to adapt more quickly to modularly varying environments by altering one module while leaving the others the same or by rearranging modules. And, as in biology, modularly varying environments are a pervasive fact of social life: human social groups often need to develop the capacity to solve new problems while retaining most of their prior problem-solving abilities. Thus, groups possessing modular mechanisms would be more likely to survive and produce "offspring" in the form of offshoot or copycat groups or organizations. However, there is reason for skepticism about this scenario.

The unit of selection in the scenario just described is the social group, and one important type of social group is the organization. In fact, there is a social science research program inspired by evolutionary biology in which the units of selection are organizations, namely, organizational ecology. Organizational ecology attempts to explain characteristics of various types of organizations—businesses, labor unions, advocacy groups, churches, and so on—in distinct contexts on the basis of differential mortality and founding rates (cf. Hannan and Freeman 1989; Aldrich 1999, 43–48). For example, one important thread in this literature examines the distinct environments to which generalist and specialist organizations are best suited, for instance, inquiring into the conditions in which consolidation among generalist organizations creates resource opportunities for specialists (cf. Carroll and Swaminathan 2000). Unfortunately, the scenario sketched in the foregoing paragraph contradicts one of the basic premises of organizational ecology: the structural inertia of organizations (Hannan and Freeman 1989, 70; Aldrich 1999, 45). According to this principle, the rate of change of an organization's structure is typically much slower than the rate of change in the environment. This premise is important for a model in which Darwinian selection is the driving force. Changes in populations of organizations result primarily from old organizations disbanding and being replaced by new ones that are better suited to the new environment rather than from individual organizations adapting themselves to new situations. There are a number of reasons why organizations would be expected to exhibit structural inertia (Hannan and Freeman 1989, 67–69). For example, restructuring often shifts resources away from a segment of the organization, and hence is likely to be resisted by those members who would be disadvantaged. Moreover, there is some empirical evidence in support of structural inertia (Aldrich 1999, 168). Thus, the proposal that highly modular, and therefore quickly changeable, organizations are favored by social selection processes is problematic.

Let us try a different approach. Modularity of social mechanisms need not entail that individual organizations be quick to adapt to changing circumstances. That point can be appreciated through a consideration of modular mechanisms in molecular biology. In that case, modularity is a matter of how the genome maps onto system components, not a claim that *individual* organisms can quickly adapt to new environments. The adaptation that modularity engenders, occurs across generations, not in the life history of a single organism. Thus, perhaps things work similarly in the social world. Consider two general ways in which this might happen.

First, consider social mechanisms that are internal to organizations. These mechanisms might include such things as a social hierarchy or an established production procedure. In this case, the argument would be that modularity facilitates evolvability because it allows mechanisms to be modified one component at a time or for solutions to new social problems to be found by rearranging mechanism components. This scenario is consistent with structural inertia, since the altered versions of the mechanism might occur in newly founded organizations rather than in transformed versions of older ones. In this scenario, nonmodular social mechanisms would be likely to go extinct in modularly varying environments, while the varied descendants of modular mechanisms would spread throughout the population of organizations. The plausibility of this scenario is enhanced by the wide prevalence of certain types of modular structures found in organizations and social groups in general, particularly hierarchies. For example, consider the hierarchical structure of a university: the university is divided into colleges or schools, which are in turn divided into departments or units. This structure is modular, since it allows alterations to be made to one unit (say, restructuring the philosophy department) while leaving other units as they were before. Likewise, although it would be difficult for an established university to, say, eliminate a number of existing departments or to restructure its colleges, a newly founded university might readily make such changes.

A second scenario concerns social mechanisms that are not internal to specific organizations, but instead are features of the broader social context in which organizations as well as individuals are embedded and interact. Forms of economic interaction, such as a market, are examples of social mechanisms of this kind. Again, the hypothesis would be that such mechanisms, if modular, are more adaptable to changing environments. As a result, such mechanisms would be expected to proliferate more widely than their nonmodular counterparts. An economic system based upon property rights and market exchange is arguably a modular mechanism, since it allows owners wide leeway to modify their properties or enterprises independently of others (cf. Langlois 2002, 26–27). Such a system also allows for rearrangement of modules in the form of consolidation or increasing specialization of industries. A more specific example is the contrast between traditional and Silicon Valley models of research and development (Aoki and Takizawa 2002). In the traditional model, R&D is carried out in an integrated manner within a particular firm, which organizes and directs coordinated R&D projects for specific goals. In this model, it is important that each of the various design teams knows what the others are doing, so that their results can be assimilated into the final product. Clearly, communication among design teams becomes increasingly cumbersome with the increasing complexity of the task of each. In the Silicon Valley model, by contrast, the product system is divided into modules developed by separate firms, often start-ups funded by venture capitalists. The Silicon Valley model requires standardized interfaces between modules, so that improvements to the overall product system result primarily from independent improvements in the various components (Aoki and Takizawa 2002, 770–71). The advantage of the Silicon Valley model is that it avoids the onerous communication among design teams required by the traditional model, thereby facilitating quicker solutions to new problems. The Silicon Valley model, then, is an example of a modular mechanism that structures the interactions of a collection of organizations. But there is nothing in this scenario to require that individual organizations be highly adaptable.

The two scenarios described above illustrate ways in which the hypothesis about the advantages of modularity with regard to evolvability might be extended to social mechanisms. But the quantity of both theoretical and empirical research on these questions in social science is minuscule in comparison to the body of work on modularity and evolvability in biology. Robert Boyd and Peter Richerson (Richerson and Boyd 2005; Boyd and Richerson 2005) are the only authors I know of who have offered anything like a detailed evolutionary explanation of modularity in social science. Boyd and Richerson argue against the image of culture as a tightly integrated, holistic system (Richerson and Boyd 2005, 91–93), and they hypothesize that culture evolved as an adaptation to rapidly changing climates in the Pleistocene (ibid., 131–39). They develop models that illustrate how the cumulative social learning indicative of culture can be favored by natural selection in changing environments (Boyd and Richerson 2005, pt. I). The main theme of this account is that culture enhances adaptability by facilitating quick, though not necessarily optimal, solutions to new problems. Hence, Boyd and Richerson's hypothesis is very similar to the evolutionary account of modularity described in section 3.4.2. Nevertheless, the focus of Boyd and Richerson's work is explaining the origin of culture rather than modularity per se, and it is unclear to what extent their proposals could be developed to support the claim that specific types of social mechanisms are modular.

In the remainder of this section, I consider some possible reasons for thinking that social mechanisms may often be nonmodular. The first concern is based on the point that modularity is adaptively beneficial only in changing environments. Consequently, nonmodular designs may be preferable to modular ones in environments that exhibit a high degree of stability over time. Thus, there would appear to be no particular reason to expect modular social mechanisms in social contexts that have persisted without much change for a significant period. Richard Langlois suggests that certain nonmodular features of medieval European social structures were well suited to the stable social environment of this period, but eventually disappeared in the face of changing circumstances (2002, 28–29). Of course, the analogous point holds with regard to biology as well. Thus, the question here is to what extent past social and biological environments have been modularly variable rather than stable or simply chaotic. The next concern, however, is more specifically focused on characteristic features of human society.

A common challenge for social policy is that changes in one feature of a society may produce unpredictable changes elsewhere in the system, thus making it extremely difficult to anticipate the consequences of the policy intervention. One source of this difficulty is that participants in the system who are not directly targeted by the policy intervention may nevertheless be aware of it, and may perceive opportunities to advance their interests by modifying their practices in response to it. Indeed, the complex interrelation between social structures and awareness of those structures by members of the society is a common basis for arguments against the possibility of laws of social science (cf. Searle 1984; Taylor 1971). Although such arguments rarely use the term "modularity," the modularity of social mechanisms is precisely what they aim to call into question. For if the objection is correct, it will typically not be possible to change one component of a social mechanism without producing unpredictable changes in the others.

This objection to the modularity of social mechanisms will be discussed in detail in Chapter 8. For the moment, I would like to indicate two points that would be relevant to any response to it. Whether a mechanism is modular with regard to an intervention depends on the intervention itself and on the manner in which the causal system is represented. For a given mechanism, some interventions may be modular while others are not. In Chapter 8, I call interventions that affect mechanisms in nonmodular ways *structure-altering*. The second point is that even if an intervention is structure-altering with regard to a mechanism, it might not be such with regard to other, more fundamental mechanisms that can explain why and how the first was altered. Thus, one natural response to the objection described in the foregoing paragraph is that the unintended effects of the policy intervention could be explained, and perhaps even anticipated, by individual-level mechanisms. For example, a rational choice model might explain why an intervention that inadvertently creates new incentives leads to systematic but unintended changes of behavior. The thought that more fundamental, modular mechanisms can be described at finer-grained levels of description is an underlying motivation of the mechanisms approach to extrapolation. It is also the central theme of Chapter 7, which discusses the relationship between mechanisms-based extrapolation and reductionism.

3.6 CONCLUSION

This chapter began with the question of the relationship between mechanisms and the probabilistic causal concepts elaborated in Chapter 2, and it proposed the first part of an answer to this question. To the extent possible, mechanisms are to be identified with causal structure on the basis of domain-specific empirical analyses. Since causal structure is that which generates probability distributions and provides information about how they change under interventions, this identification is a basis for linking mechanisms to probabilistic causal concepts. An important part of these empirical analyses consists of providing some general reason to think that mechanisms are modular, and evolutionary theory suggests a means of doing just this. However, this evolutionary argument is, at present, on firmer ground in molecular biology than in social science.

Yet the identification of mechanisms with causal structure alone indicates only that there is *some* connection between mechanisms and probabilistic causal concepts such as causal effect and positive causal relevance. It provides no indication of what the nature of that relationship is. Chapter 4 discusses a proposition, which I call the *disruption principle,* which says something specific about the link between probability and mechanisms identified with causal structure.

4

The Disruption Principle

All of the known mechanisms by which HIV impairs the human immune system depend on HIV reproduction. Therefore, the development of anti-HIV replication drugs would appear to be a positive first step in controlling HIV reproduction.

—Gerald Stine, *AIDS Update 2000* (2000, 84)

Implicit in the above quotation is a commonsense idea that I think is widespread in much of biology and probably in many other areas of science as well: a causal effect is completely nullified when, and only when, every mechanism linking cause and effect is severed. This is a rough statement of what I call the *disruption principle*, which is the main topic of this chapter. The disruption principle specifies a relationship between mechanisms, identified with causal structure as proposed in Chapter 3, and probabilistic causal concepts such as causal effect and causal relevance. In virtue of making this connection, the principle plays an important role in the treatment of extrapolation developed in subsequent chapters. The purpose of this chapter is to articulate the disruption principle, illustrate it by reference to a scientific example, and explore the range of circumstances in which it can be reasonably presumed.

Since a concrete example greatly facilitates the presentation of the disruption principle, I begin with a brief description of the mechanism of HIV replication, and of some of the factors known to interfere with it. I then turn to a statement of the principle itself. Formulating the disruption principle in a precise way involves developing a graphical framework for representing factors that disrupt mechanisms, so this is somewhat complex. I then illustrate the disruption principle and the graphical framework in question by means of an example drawn from HIV research, namely, the discovery of a genetic mutation that confers substantial resistance to HIV infection. In the abstract, the problem illustrated by this example takes the following form. Suppose we know that a certain causal relationship holds between X and Y in the population P, for example, that X is a positive causal factor for Y. We want to know if there is a subpopulation of P in which this effect is nullified. From the disruption principle it obviously follows that such a population must be one in which every mechanism from cause to effect is blocked. The challenge, then, lies in ascertaining whether such a subpopulation exists, given a lack of full knowledge of the total set of mechanisms, a state of incomplete knowledge common in biological and biomedical contexts.

Given the important role of the disruption principle, it is worthwhile to consider what justification there is for assuming it. I show that, given the identification of causal structure and mechanisms argued for in Chapter 3, the disruption principle can be shown to follow from the conjunction of two more familiar principles concerning causality and probability: the principle of the common cause (PCC) and the faithfulness condition. This suggests that the disruption principle might be false—and hence the mechanisms approach to extrapolation, unreliable—when one or both of these principles do not obtain. That point is hardly trivial, since doubts have been raised concerning both the PCC and the faithfulness condition (cf. Sober 2001; Cartwright 1999). I argue that the PCC is on firm ground with respect to the types of cases that concern the disruption principle. The situation in the case of the faithfulness condition, in contrast, is more complex. I suggest that there is a strong motivation for the FC when studying heterogeneous populations, but that this justification collapses for exceedingly homogeneous populations, such as closely inbred strains of laboratory mice reared under uniform conditions. This result entails that special care should be taken to vary genetic or environmental background conditions in gene knockout experiments, a point which some researchers in this field have noted. That heterogeneity can be a virtue in scientific experiment is surprising in light of the common notion that the ideal experiment is one in which all factors except those subject to investigation are held constant.

4.1 HIV REPLICATION

According to the current standard, AIDS is diagnosed when a person's T-helper cell count drops below 200 per microliter of blood (cf. Stine 2000, 132; Kalichman 1998, 78–79).[1] Hence, the role of T-helper cells in the human immune system is a good place to begin in describing HIV replication.

The primary actors of the immune system are a collection of distinct types of white blood cells that identify and destroy antigens present in the body. For example, phagocytes eat bacteria and foreign or infected cells, while mast cells and eosinophils attack intruders too big for consumption (e.g., worms) by emitting poisonous chemicals in their vicinity (cf. Fan et al. 2000, 26–27). However, the cells of greatest concern for our purposes are lymphocytes. These come in two basic varieties, the B lymphocytes and the T lymphocytes. T lymphocytes themselves come in several varieties, the most important of which for our purposes are T-helpers and cytotoxic T-cells, or T-killers. The B lymphocytes identify which entities in the body are to be attacked by phagocytes, mast cells, and eosinophils (ibid., 32–37). B lymphocytes perform this function by producing antibodies, which are proteins that attach to particular sorts of intruding agents. Different B lymphocytes produce different antibodies, the specific antibody produced being determined by the result of a random rearrangement of DNA within the cell. Once mature, a B lymphocyte will not replicate itself or emit its antibodies into the bloodstream unless two things happen. First, it must encounter an antigen to which its antibody attaches. Next, it must encounter a T-helper cell that also attaches to this antigen; when this happens, the T-helper chemically signals the B lymphocyte to release its antibodies and commence replicating. Hence, in the absence of T-helpers, the immune system is unable to identify which bodies are to be destroyed (e.g., by phagocytes) and which are not.

T-killer cells differ from B lymphocytes in that they directly attack and destroy antigens, yet in other respects the functioning of the two types of cells are very similar (cf. Fan et al. 2000, 42–47). Like B lymphocytes, different T-killer cells have a chemical affinity for different types of antigens. Moreover, a T-killer cell that has encountered an antigen to which it attaches will not replicate until instructed to do so by a T-helper cell that recognizes the same antigen. Thus, like the B lymphocytes, the T-killer cells can perform their role within the immune system only in the presence of T-helper cells. It can easily be understood, then, how a large-scale reduction in the number of T-helper cells would lead to catastrophic failure of the immune system and to opportunistic infections.

Given this background, we can proceed to a description of the mechanism by which HIV infects and destroys T-helper cells. It should be noted that T-helper cells are not the only cells of the human immune system that are infected by HIV. For example, a type of phagocyte, namely the macrophage, is also prone to HIV infection. Indeed, in the early and nonsymptomatic stages, HIV infection is restricted almost exclusively to macrophage-tropic (M-tropic) HIV, with widespread infection of T-helper cells occurring later in the progression of the disease (Zhu et al. 1993).[2] Macrophages will play a significant part in the discussion in the following section. We will see there that a few lucky individuals possess a genetic mutation that blocks HIV entry into macrophages, thereby conferring a high degree of resistance to HIV infection. Nevertheless, the mechanism of HIV replication is typically presented in textbooks at a level of abstraction that does not distinguish between the two cases (cf. Stine 2000, 64; Kalichman 1998, 16; Fan et al. 2000, 59). Let us turn now to a description of this mechanism.

HIV is an example of a retrovirus, which is so called because it reverses the normal flow of information from DNA to RNA. HIV replication proceeds according to the usual pattern for retroviruses. The genetic material of a retrovirus is encoded by RNA, and when a retrovirus infects a cell, its RNA serves as a template for the transcription of viral DNA, which is then insinuated into the cell's nuclear DNA. Once this occurs, the cell becomes a factory that produces HIV. The viral DNA integrated into the cell's genetic material codes for new viral RNA and proteins necessary for the functioning of the retrovirus. These materials are then assembled in the cytoplasm and new retroviruses bud from the cell membrane, ultimately destroying the cell if they do so in sufficiently large numbers.

The mechanism by which HIV infects T-helper cells and macrophages is often depicted by diagrams like that in Figure 4.1 (cf. Kalichman 1998, 16; Stine 2000, 64). With the aid of such a diagram, it is possible to introduce more details about the mechanism than those just sketched above (cf. Kalichman 1998, 15–17; Stine 2000, Chapter 3). Glycoproteins (gp120) protruding from the surface of the HIV retrovirus attach to the T-helper cell at the CD4 (cluster determinant-4) receptor site. Note, therefore, that HIV infects only cells, such as macrophages and T-helpers, which display the CD4 receptor on their outer surface. Next, the viral RNA, ensconced in a protein coat, is injected into the host cytoplasm. Along with RNA, several enzymes necessary for the continuation of the infection are contained within the protein coat—most prominently, reverse transcriptase and integrase. The protein coat is quickly dissolved, and the viral DNA is then transcribed from the viral RNA by means of reverse transcriptase. The viral DNA is then integrated into the DNA of the host cell with the aid of integrase.

The cellular machinery of the host then proceeds to transcribe viral RNA and to synthesize proteins from the intruding DNA, thereby generating the materials needed to create new HIV viruses. These materials include new strands of viral RNA as well as several proteins and enzymes necessary for the functioning of the retrovirus. Besides reverse transcriptase and integrase, the enzyme protease is produced at this stage. Protease performs the role of splicing long protein strands into small, more usable pieces from which the internal protein coat can be constructed. Finally, the materials that constitute the new HIV virus assemble near the cell's external border, taking part of the host cell membrane with them as they bud forth. A large number of budding HIV viruses, therefore, kill the host

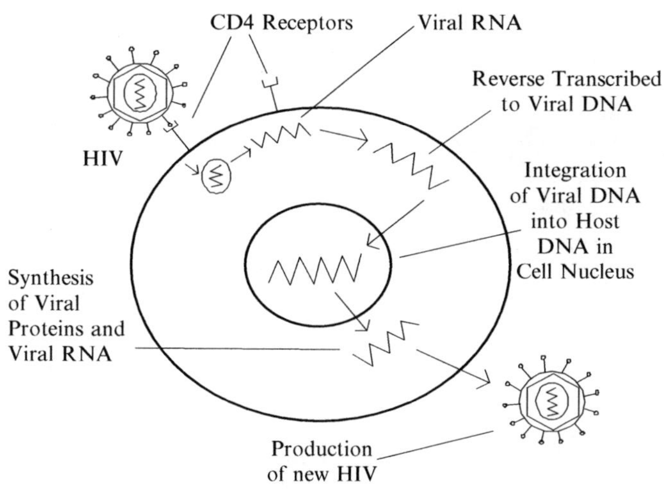

Figure 4.1 HIV replication

cell by creating numerous perforations in its membrane.[3] Shortly after exiting the host cell, the virus constructs its inner structure of protein layers that enclose the RNA and vital enzymes. It is now ready to infect a new host.

Such is the mechanism by which HIV replicates. The evolutionary argument for why mechanisms can be identified with causal structure proposed in section 3.4 can be easily applied in this case. The HIV replication mechanism has clearly been honed by natural selection, which maintains the "normal" pattern described above as the statistically typical one. Moreover, the exasperating ability of HIV to evolve resistance to anti-retroviral therapies is a clear signal of the modularity of HIV replication. Thus, there is every reason to think that the HIV replication mechanism is capable of generating probability distribution and providing information concerning how those distributions will change, given interventions. In short, it is a causal structure.

The HIV replication mechanism depicted in Figure 4.1 is also easily represented by a directed graph. For example, consider a collection of binary variables (1 = yes, 0 = no) defined as follows:

X: exposure to HIV
A: the virus attaches to the CD4 receptor
B: viral material enters the cytoplasm
C: reverse transcription of viral RNA occurs
D: viral DNA is integrated into host DNA
E: viral materials are produced by host cell
F: viral materials are assembled in preparation to form a new HIV virus
Y: a new infectious virus buds from the cell

Then we can represent the HIV replication mechanism with the graph in Figure 4.2. Of course, the mechanism could be represented in more or less detail, but this will suffice for present purposes.

4.2 FORMULATING THE PRINCIPLE

The disruption principle serves as a bridge from knowledge of mechanisms and things that interfere with them to qualitative conclusions about causal effects. Imagine that *X* is a positive causal factor for *Y* in the population P. Now suppose we ask whether there is a proper subset P′ of P in which the effect of *X* upon *Y* is *nullified*, that is, such that *X* is not causally relevant to *Y* within P′. The disruption principle provides a necessary and sufficient condition for the existence of such a subpopulation: for each member of P′; every mechanism from *X* to *Y* is blocked.

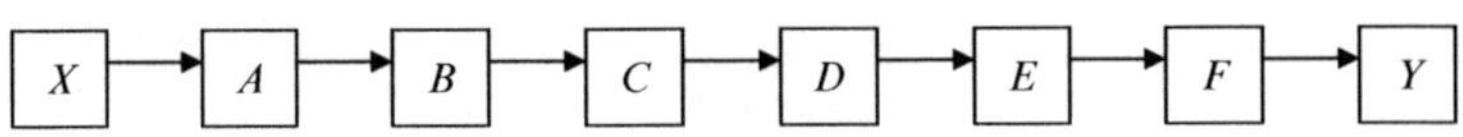

Figure 4.2 A directed graph representing HIV replication

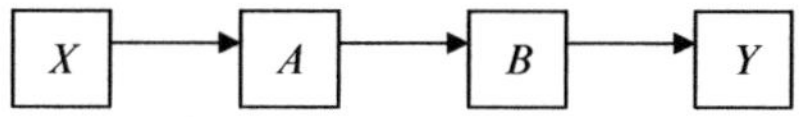

Figure 4.3 The flashlight mechanism

The disruption principle entails that if interventions on X alter the probability distribution of Y, there is at least one mechanism from X to Y. I will assume that directed graphs represent mechanisms as paths in which all of the arrows point in the same direction, as in the graph for the HIV replication mechanism given in Figure 4.2. Since that graph was rather lengthy, it will be convenient to use a shorter one for the purposes of illustration. For example, consider the simple case of pushing the "on" button of a flashlight. Pushing the button closes the electrical circuit, causing electricity to flow to the bulb, which in turn lights up. This causal chain could be represented by the graph in Figure 4.3, where X, A, B, and Y are binary variables representing, respectively, the button being pushed, the circuit being closed, electricity reaching the bulb, and the light shining. Thus the graph in Figure 4.3 depicts the mechanism of the flashlight.

A factor that nullifies the effect of X upon Y, then, must break this causal chain at some point. In the present example, such a factor is ready at hand; namely, the battery being dead. Let the variable Z represent the state of the battery: $Z = 1$ if the battery is charged and 0 otherwise. Adding Z to the graph from Figure 4.3, we have Figure 4.4. Notice that this graph does not indicate that X and Z interactively influence Y. That is, when $Z = 0$, X has no influence on Y; but this information is omitted by the graph in Figure 4.4. Indeed, the graph in Figure 4.4 could represent a situation in which A and Z influenced B independently of one another.[4]

Stating the disruption principle, then, is aided by a graphical notation for representing causal interactions in which an interfering factor severs a mechanism. I shall presume that each mechanism is represented by one directed path, such as that from X to Y in Figure 4.4. This is a purely terminological decision made for the pragmatic reason that it facilitates stating the disruption principle. Thus, several related causal chains that might be naturally referred to as a single mechanism would, in my terminology, be described as several interacting mechanisms. Factors that disrupt a mechanism, then, can be represented as follows. In the above example, there is a range of values of the variable Z (in this case

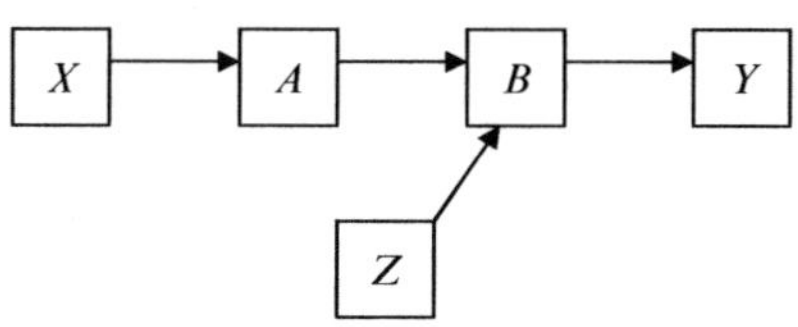

Figure 4.4 The flashlight mechanism and the battery

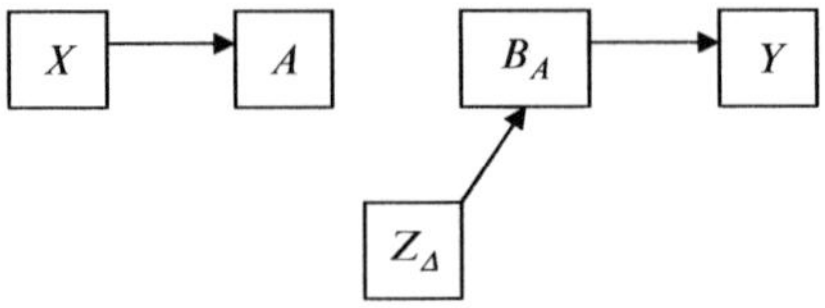

Figure 4.5 A disrupting factor

{0}) such that, when the value of Z is in that range, the effect of A upon B is nullified. I represent the elimination of the causal relationship between these variables by deleting the arrow joining them. Let Z_Δ indicate that the value of Z is such as to break the causal chain (the subscript delta standing for "disrupt"). In the example just described, Z_Δ would indicate that the battery is dead (i.e., $Z = 0$). I shall refer to variables with "Δ" subscripts as *disrupting factors*. Then the causal structure for the subpopulation composed entirely of otherwise functional flashlights with dead batteries can be represented by the graph in Figure 4.5. The subscript A appended to B indicates that there is an arrow from A to B in the graph representing the causal relationships in the general population. Thus, given the graph in Figure 4.5, it is possible to unambiguously reconstruct the graph representing the causal relationships that hold in the general population, wherein the value of Z is not restricted to the disrupting set of values.

A disrupting factor, then, can be thought of as a switch that, when set to a particular position, breaks the mechanism connecting the cause and the effect. A more exact definition can be provided as follows. I shall use the expression *precedent variable* of a given node on the mechanism to refer to the directly prior node. For instance, in the mechanism represented in Figure 4.3, A is the precedent variable on B. Suppose that M is a mechanism through which X influences Y. Let V ($\neq$ X) be a variable on M. Then

> *Definition 4.1*: Z is a *disrupting factor with respect to M* just in case there is a variable V in M such that (1) Z is a cause of V, and (2) there is a range or interval Δ of values of Z such that when the value of Z is in Δ, the variable in M precedent to V is not a direct cause of V.

For example, in Figure 4.4, Z is a cause of B, and when $Z = 0$, the variable on the mechanism precedent to B—namely, A—is not a cause of B. A disrupting factor Z will be said to be *active* with respect to a particular individual p if the value of Z for p is in the interval or range Δ that results in the mechanism being disrupted.

Since there may be several mechanisms connecting a given cause and effect in a population, it will be useful to speak of a *mechanism set* for a pair of variables. The mechanism set from X to Y in a population P consists of all of the mechanisms through which X influences Y that are found in at least one member of P. I shall use the notation $\mathbf{M}_{XY}$ to represent the mechanism set from X to Y. Mechanism sets are hence relative to a population. I shall use the expression "$\mathbf{M}_{XY}$ for p" to designate the subset of $\mathbf{M}_{XY}$ that is instantiated in the individual p, where p is any member of the population P of concern. Then:

> *Definition 4.2*: $\mathbf{M}_{XY}$ for p is *disrupted* if and only if, for each M in $\mathbf{M}_{XY}$ for p, there is at least one disrupting factor that is active with respect to p.

Notice that, given definition 4.2, if $\mathbf{M}_{XY}$ for p is empty, then it is trivial that $\mathbf{M}_{XY}$ for p is disrupted. Let φ_0 be the relative frequency of individuals in P for which $\mathbf{M}_{XY}$ is disrupted. Then:

> *Disruption principle*: X is causally relevant to Y in P if and only if $\varphi_0 < 1$.

The disruption principle, therefore, links mechanisms to the probabilistic concept of causal relevance from Chapter 2. It may be helpful to quickly retrace these steps. According to the Machamer-Darden-Craver definition, "Mechanisms are entities and activities organized such that they are productive of regular changes from start or set-up to finish or termination conditions" (2000, 3). Chapter 3 argued that mechanisms so defined can, at least within the domain of molecular biology, be identified with causal structure. Causal structure, meanwhile, is defined as that which generates probability distributions and provides information concerning how those distributions change given interventions. Directed graphs, in turn, are one useful means of representing causal structure, and hence mechanisms if the two are identified. Finally, the disruption principle uses a slightly modified directed graph framework to state a relationship between mechanisms and the probabilistic concept of causal relevance defined in section 2.3.

Recall that X is causally relevant to Y just in case ideal interventions on X make a difference to the probability distribution of Y. As specified in definition 2.1, an ideal intervention is an exogenous cause that determines the value of the variable it targets. In the flashlight example, this could be something as simple as pushing the switch to "on." So, suppose our population P consists solely of flashlights with dead batteries. Then the graph in Figure 4.5 represents the disrupted mechanism found in each member of P, which means that $\varphi_0 = 1$. Therefore, by the disruption principle, X is not causally relevant to Y in P, that is, $P(Y = 1 \mid do(X = 1)) = P(Y = 1 \mid do(X = 0))$. In other words, when the battery is dead, moving the switch from "on" to "off" makes no difference to the probability that the light is shining. On the other hand, if there are some flashlights in P that possess the undisrupted mechanism, then $\varphi_0 < 1$, and hence the disruption principle entails that $P(Y = 1 \mid do(X = 1))$ is not equal to $P(Y = 1 \mid do(X = 0))$. Since $P(Y = 1 \mid do(X = 0)) = 0$, this entails that X is a positive causal factor for Y, that is, $P(Y = 1 \mid do(X = 1)) > P(Y = 1 \mid do(X = 0))$. The disruption principle, therefore, can link mechanisms to claims about positive causal relevance.

It should be noted that nullifying the effect of X upon Y does not necessarily entail determining the value of Y. For example, imagine there is a gene that neutralizes the effect of smoking upon lung cancer. In the population of people possessing this gene, then, smoking would not be causally relevant to lung cancer. Nevertheless, members of this population might develop the disease through exposure to other carcinogens. On the other hand, neutralizing the effect of HIV in a population is sufficient for ensuring that nobody develops AIDS. Likewise, if the battery of the flashlight is dead, then the light does not shine. It is important to bear in mind, then, that this is a special feature of these two examples. It is not the case in general that eliminating the effect of *X* upon *Y* determines *Y*'s value.

Of course, the disruption principle is of little use unless some knowledge of relevant mechanisms and disrupting factors is available. Given the identification of mechanisms with causal structure, learning about mechanisms can be viewed as a special case of causal inference more generally conceived.[5] There are some discussions in the philosophical literature that examine strategies specifically suited for learning about mechanisms (cf. Bechtel and Richardson 1993; Darden 1991, 2002; Darden and Craver 2001, 2002). One such strategy, known as process tracing, is described in Chapter 5. Chapter 9 examines the relationship between process tracing and causal inference from statistical data. But for the moment, I set aside the concerns about how knowledge of mechanisms is to be acquired, assuming that the inquiry commences with some information on this score. Given such knowledge, the hunt for interfering factors can proceed by identifying points at which the mechanism is vulnerable to interference and searching for variables in the population capable of interfering with the mechanism at the specified points. Our knowledge of the mechanism need not be perfect for this hunt to commence, and as the example in the following section illustrates, the search for interfering factors can itself result in significant improvements in our knowledge of a mechanism.

4.3 RESISTANCE TO HIV INFECTION

Let us examine how the disruption principle comes into play in a realistic scientific example. Consider the question of whether there is a subpopulation in which the effect of exposure to HIV upon AIDS is nullified. It might seem that there is a straightforward solution to this problem that is independent of mechanisms: one need only find those who have been exposed to HIV but have not become infected. However, such a method, on its own, is an unreliable means for discovering subpopulations in which a causal effect is nullified, since there are several possible explanations for why the effect might not have followed the cause in a given case, including pure luck, exposure to a very mild form of the virus, or intrinsic resistance. The first two of these explanations yield very different predictions than the third about how the individual would respond to future exposures.

The disruption principle tells us that a fully resistant subpopulation, if it exists, is one in which every mechanism from HIV exposure to AIDS is severed in each individual. Hence, given the disruption principle, the search for the subpopulation in which the effect of HIV exposure upon AIDS is eradicated becomes the search for a disrupting factor, or set of disrupting factors, capable of blocking all mechanisms through which HIV brings about AIDS. Since all such mechanisms depend on HIV replication, that mechanism is a good place to look for such disrupters. That is, the set of mechanisms through which HIV produces the suite of symptoms associated with AIDS can be thought of as having the shape of a fan, with replication as its stem. Thus, since each mechanism shares this stem, blocking replication would sever all of them in one fell swoop. This thought is the point of the quotation at the head of this chapter. Let us turn, then, to the story of the discovery of a disrupting factor that seemed capable of nullifying the effect of HIV.

As described in section 4.1, HIV replication begins with the HIV retrovirus attaching to the CD4 receptor, which is exhibited on the surface of T-helper cells and cells of several other types, such as macrophages. However, the presence of the CD4 receptor is generally sufficient for an HIV virus to attach to a cell but not sufficient for the entry of the viral core into the cytoplasm (Maddon et al. 1986). Moreover, HIV strains that are capable of infecting macrophages are generally not able to infect noncirculating T-helper cells found in lymph nodes, and vice versa (cf. Gartner et al. 1986; Stine 2000, 141). Although these facts were recognized within a few years of the discovery of HIV,[6] an explanation of them was not forthcoming until nearly a decade later.

In 1996, it was discovered that distinct co-receptors present on macrophage and noncirculating T-helper cells play an important role in the entry of viral material into the host cell (Deng et al. 1996; Dragic et al. 1996). The co-receptor in the case of noncirculating T-helper cells is called CXCKR4 (X4 for brevity), and its counterpart for macrophages is known as CC-CKR5 (R5 for brevity).[7] M-tropic HIV utilizes R5, while T-tropic HIV depends upon X4, thereby accounting for the difference in affinities of the two strains.[8] Within the same year, it was discovered that some individuals who had not become infected with HIV despite repeated exposures possessed a mutation that inhibited the normal R5 co-receptor (Samson et al. 1996; Liu et al. 1996). When exposed in vitro to M-tropic HIV strains, cells from these individuals

> …required about 1000-fold more virus to establish infection than control cells from unexposed donors. While a small fraction of the cells did become infected with this high inoculum, the virus failed to replicate further. (Liu et al. 1996, 367)

As noted in the foregoing section, M-tropic HIV predominates in the early and asymptomatic stages of infection. Thus, if replication of M-tropic HIV is blocked, the progression of the infection is strongly, if not completely, inhibited. The absence of the normal R5 co-receptor was linked to a homozygous mutation, in which thirty-two base pairs in the ordinary gene coding for the co-receptor were deleted. As this mutation appears to produce no other abnormal phenotypic effect, it is a veritable genetic blessing for those lucky enough to have inherited it. The homozygous mutation was estimated to occur among approximately 1 percent of "persons with western European heritage" (Liu et al. 1996, 373; Dean et al. 1996). The heterozygous condition, which appears to confer a more attenuated resistance (Eugen-Olsen et al. 1996), is surprisingly common—with estimates ranging from about 20 percent (Liu et al. 1996, 373) to 14 percent (Dean et al. 1996, 1860) among Caucasians. The mutant allele was not found in African or Asian populations (Samson et al. 1996, 722). There was a striking negative association between HIV infection and the homozygous mutation. In several large data sets, *all* of those homozygous for the thirty-two-base-pair deletion were HIV negative (Samson et al. 1996, 722; Dean et al. 1996, 1860). These data stimulated hope that the homozygous mutation affecting the R5 co-receptor might confer complete resistance to AIDS.[9]

The thread of this scientific detective story will be taken up again in Chapter 7, so for now let us consider what, if the disruption principle is true, would have to be the case for the hope just described to be realized. Consider the segment of the M-tropic HIV replication mechanism that is disrupted by the homozygous mutation affecting the R5 co-receptor, which is represented in Figure 4.6.

As in Figure 4.2, *X* and *A* are binary variables indicating exposure to HIV and attachment of HIV to the CD4 receptor, respectively, while *R* is a binary variable indicating attachment to the R5 co-receptor and *V* represents the rate of reproduction of M-tropic HIV. Suppose that the presence of the homozygous thirty-two-base-pair deletion fully blocks attachment to the R5 co-receptor. Then we have Figure 4.7. Here *Z* is a variable representing the presence of the mutation affecting the R5 co-receptor that takes three values {homozygous normal; heterozygous; homozygous mutant}. The subscript "Δ" in this case indicates that *Z* takes on the third of these values. Thus, given the supposition that the homozygous mutation completely blocks the mechanism in Figure 4.6, it also fully blocks M-tropic HIV reproduction if there is no path from *X* to *V* that circumvents *R*.

But even if this is so, it would not necessarily follow that the homozygous mutation confers immunity to HIV infection and AIDS, since T-tropic HIV does not utilize the R5 co-receptor. However, given that M-tropic HIV predominates in the early stages of HIV infection, it is possible that the continuation of infection by T-tropic HIV depends

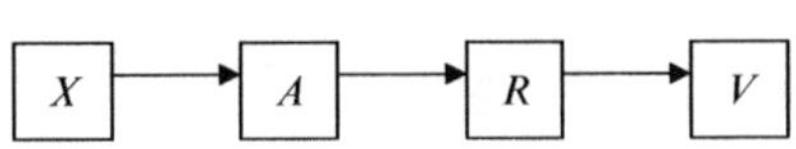

Figure 4.6 M-tropic HIV replication

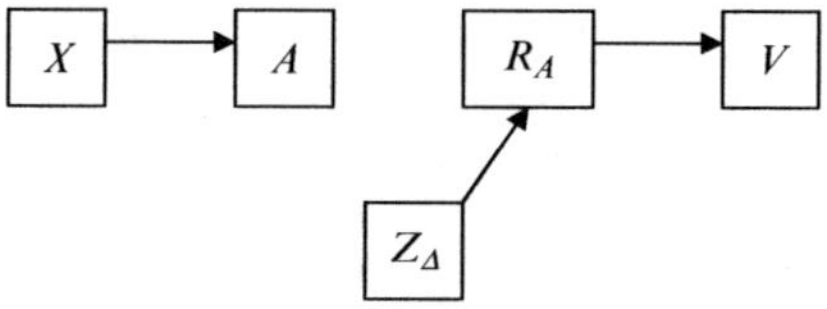

Figure 4.7 The homozygous mutation

upon the replication of M-tropic strains. Letting V_{Δ} indicate $V = 0$, this thought can be represented by the graph in Figure 4.8. In this graph, S is a binary variable indicating entry to the cytoplasm for T-tropic HIV, and T is some unspecified stage of T-tropic HIV replication that is blocked by the failure of the M-tropic HIV infection. Precisely what T might consist of depends on how the absence of M-tropic replication inhibits that of T-tropic, an issue that will be taken up in section 7.1.

From the graph in Figure 4.8, it can easily be seen that, given the disruption principle, the homozygous mutation inhibiting the R5 co-receptor completely nullifies the effect of HIV exposure upon AIDS if and only if there is no path from HIV exposure to AIDS that bypasses both R and T. In other words, it must be that the homozygous mutation completely blocks the M-tropic HIV replication mechanism (or the set of them, if there are several), *and* there is no mechanism from HIV exposure to AIDS that bypasses replication of M-tropic HIV. If it were to be discovered that the mutation inhibiting the R5 co-receptor did *not* confer complete immunity, the disruption principle would entail that at least one of these two conditions is false.

This example illustrates how the disruption principle captures a relatively commonsense inference concerning mechanisms and nullified causal effects. But if the role of the disruption principle were limited to reconstructing such examples, there would hardly seem to be much point in taking the time to provide a precise articulation of it. However, there is a twofold value in clearly stating and highlighting the disruption principle. First, as will be seen in Chapter 6, there are less obvious consequences of the disruption principle regarding extrapolation of causal

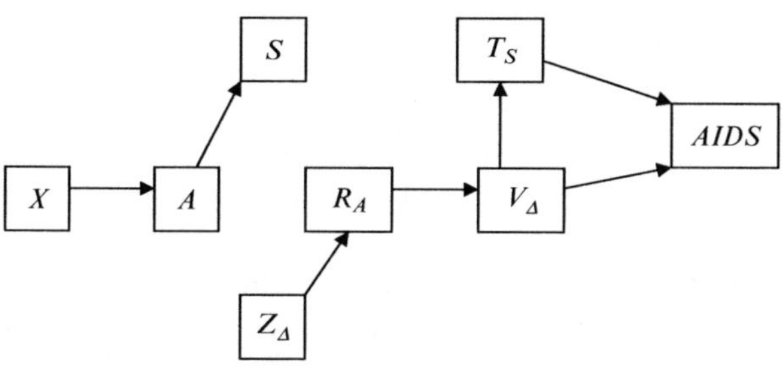

Figure 4.8 How the homozygous mutation might confer HIV immunity

claims. Second, a clear statement of the principle facilitates a careful examination of the circumstances in which it is and is not a reasonable assumption, a topic addressed in the remainder of this chapter.

4.4 WHY BELIEVE THE DISRUPTION PRINCIPLE?

I hope that the plausibility of the disruption principle has been motivated by the HIV example examined above. Nevertheless, it is worth considering whether the principle is supported by other more general, and more familiar, principles connecting causal structure and probability. In fact, granting that mechanisms are identified with causal structure, the disruption principle is a straightforward consequence of two propositions known as the *Principle of the Common Cause* (PCC) and the *Faithfulness Condition* (FC). This result is of interest in two respects. First, it implies that any justification for the PCC and the FC is also a justification for the disruption principle. Second, it suggests that circumstances in which the PCC or FC fails may be ones in which the disruption principle fails as well. Since alleged counterexamples have been raised against both the PCC and the FC, this last observation is far from being an idle point. However, I show that the PCC is on very firm ground in the type of experimental context that is of concern here. The case of the FC is more complex. I argue that the FC is reasonable for heterogeneous populations—such as naturally occurring biological populations—but not necessarily for extremely homogeneous ones, such as closely inbred strains of laboratory mice.

4.4.1 The Disruption Principle and the PCC

Let us begin by considering the connection between the disruption principle and the PCC.[10] The PCC can be stated in the following way.

> *PCC:* For any two distinct variables X and Y, if X and Y are not causally connected, then they are probabilistically independent.

Two variables are *causally connected* just in case one is a cause of the other or there is a common cause of both. Thus, the PCC says that two variables are probabilistically dependent only if one is a cause of the other or there is a third variable that is a common cause of both.

Given the identification of mechanisms with causal structure defended in Chapter 3, one half of the disruption principle is a direct consequence of the PCC. That is, the disruption principle is a biconditional that, in one direction, asserts: If there is no undisrupted mechanism from X to Y in the population P, then X is not causally relevant to Y in P. By definition 2.3, X is causally relevant to Y just in case X and Y are probabilistically dependent under ideal interventions on X. But in the context of an ideal intervention on X, Y does not cause X and there is no common cause of X and Y. Thus, if there is no undisrupted mechanism from X to Y, then X and Y are not causally connected, given an ideal intervention on X. From

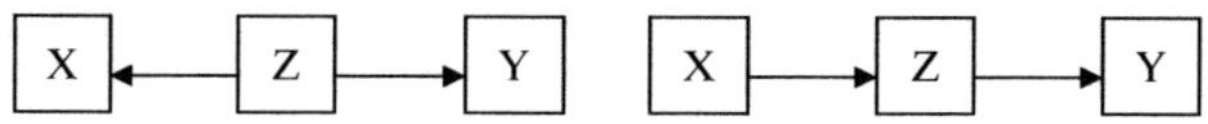

Figure 4.9 An illustration of the causal Markov condition

PCC, therefore, it follows that *X* and *Y* are probabilistically independent in such circumstances.

The PCC is itself a consequence of a more general principle connecting causality and probability known as the *causal Markov condition* (CMC). Roughly, the CMC asserts that, conditional on its direct causes, any variable is probabilistically independent of any set of other variables that do not include its effects. The CMC, therefore, entails the familiar "screening-off" rule. For example, consider the two directed graphs in Figure 4.9. If these graphs satisfy the CMC, then *X* and *Y* are probabilistically independent, conditional on *Z* in both.

Probably the most common basis provided for the CMC is that it is true of acyclic, deterministic causal structures in which the exogenous variables are probabilistically independent (cf. Pearl 2000, 30; Spirtes, Glymour, and Scheines 2000, 32; Glymour 2001, 27).[11] This proposition can be extended to indeterministic causal structures (Steel 2005), leaving only the other two assumptions—probabilistic independence of exogenous variables and absence of causal cycles—as matters of concern. The disruption principle asserts, in part, that if there are no undisrupted mechanisms from *X* to *Y*, then ideal interventions on *X* make no difference to the probability of *Y*. Recall that an ideal intervention is exogenous, that is, it is neither an effect of, nor shares a common cause with, any of the variables being studied (in this case, *X* and *Y*). The best way to ensure that this condition is satisfied in practice is to assign the value of the targeted variable (*X*, in this case) on the basis of some random process, such as tossing a coin.

So, consider the relationship between *X* and *Y*, when the values of *X* are randomly assigned by an ideal intervention, which is represented by the variable *I*. It is easy to show that, in the causal structure relating only *I*, *X*, and *Y*, all exogenous variables are probabilistically independent of one another *and* there are no causal cycles. From items (a) and (b) of definition 2.1, we know that *I* is the sole cause of *X* and a direct cause only of *X*. Moreover, item (c) of definition 2.1 asserts that *I* is exogenous. Consequently, the only two possible causal structures relating *I*, *X*, and *Y* are those represented in Figure 4.10. Since there are no causal cycles in either case, the requirement that the causal structure be acyclic is satisfied. It is trivial that every exogenous variable is probabilistically independent of every other in the graph on the left, since in that graph there is only one exogenous variable, *I*. In the graph on the right, there are two exogenous variables, *I* and *Y*. But given randomization, we know that the intervention *I* is probabilistically independent of every other exogenous variable. Hence, *I* and *Y* are probabilistically independent in that graph. Thus, in

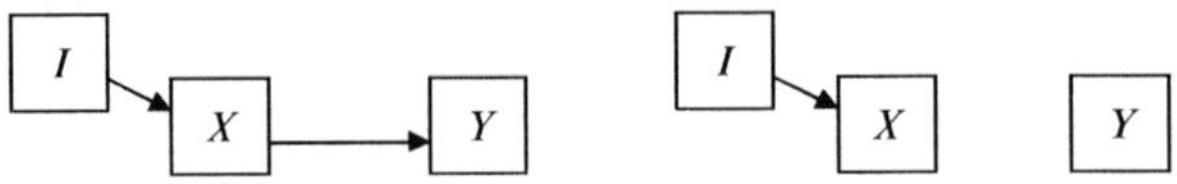

Figure 4.10 Two ideal interventions

both graphs the exogenous variables are probabilistically independent. Consequently, it follows that the CMC, and hence the PCC, is true in any case involving two variables, one of whose values is randomly assigned by an ideal intervention. Since randomization is the standard experimental procedure for ensuring that an intervention is exogenous, the half of the disruption principle asserting that a causal effect is nullified when all mechanisms are blocked is on firm ground in the context of experiments.[12]

Of course, this does not show that there are no practical challenges confronting applications of the PCC in the present context. For example, the statistical problem of reliably drawing inferences concerning probabilities on the basis of data in a sample is ubiquitous. The presence or absence of a statistically significant correlation coefficient in the data may be the result of mere chance. In addition, it may be difficult to know whether an actual experiment satisfied the conditions of an ideal intervention. But although they are real, these challenges are independent of the PCC; they are general problems for statistical inference and experiment.[13]

4.4.2 Genetic Redundancy and the Faithfulness Condition

Consider the relationship between exercise and weight loss. Additional exercise results in more calories being burned, but it also stimulates one's appetite. Letting the variables *E*, *A*, and *W* denote exercise, appetite, and weight, respectively, this situation can be represented by the graph in Figure 4.11. Conceivably, the strengths of these two paths from exercise to weight could exactly cancel out and make *E* and *W* probabilistically independent, thereby contradicting the FC. Nevertheless, it is clear that modern medicine does not take this possibility seriously: physicians and others endeavoring to promote public health have long encouraged overweight people to get more exercise.

Peter Spirtes, Clark Glymour, and Richard Scheines (hereafter, SGS) prove that though the exact balancing of strengths of counteracting causal

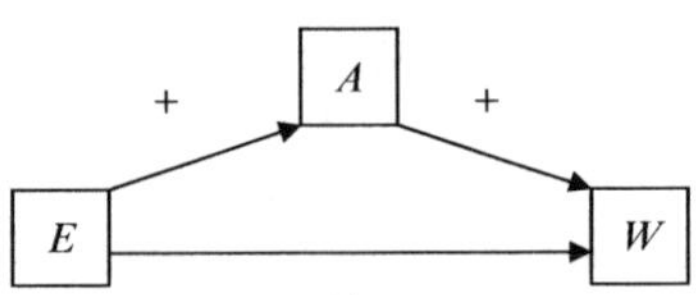

Figure 4.11 Counteracting causal paths from exercise to weight

paths is conceivable, given certain apparently plausible assumptions, it is monstrously improbable. To see the idea, consider the following linear causal model. In this model, a, b, and c are linear coefficients representing quantitative strength of influence. The subscript, lowercase e's are called "error terms," and represent any source of variation in the dependent variable not accounted for by its direct causes. It is assumed that error terms are independent and are normally distributed with zero means. A *parameterization* of a linear causal model consists in specifying numerical values for the coefficients and for the variances of the error terms. Notice that X and Z would be uncorrelated, and the FC would be false, for any parameterization in which $b + ac = 0$.

$$X = e_x$$
$$Y = aX + e_y$$
$$Z = bX + cY + e_z$$

SGS's theorem states conditions under which parameterizations that result in such precise canceling out have probability zero.[14] Take any model in which effects are linear functions of their causes. Suppose that this model contains n parameters. For example, $n = 6$ in the linear causal model. Consider the n-dimensional space of all parameterizations of this model, that is, each point in the space corresponds to a parameterization. Now consider any subset of that space consisting solely of parameterizations that violate the FC. In the model, an example of such a subset would be one in which every parameterization makes $b + ac = 0$. Then it can be shown that any subset of the n-dimensional space containing *only* parameterizations that violate the FC is of $n - 1$ dimensionality or less. Then the following assumption is made:

L: In an n-dimensional space of parameterizations, any subset of $n-1$ dimensionality or less has probability zero.[15]

Thus, it follows that any subset of the space of parameterizations of a linear causal model containing only parameterizations that violate the FC has zero probability.

However, not everyone regards SGS's theorem as a compelling motivation for the FC, and some have argued that exceptions to the FC are not uncommon. For example, according to Cartwright:

> Faithfulness will be violated if the two processes are equally effective and cancel each other out. It is not uncommon for advocates of DAG-techniques to argue that cases of cancellation will be extremely rare, rare enough to count as non-existent. That seems to me unlikely, both in the engineered devices that are sometimes used to illustrate the techniques and in the economic and medical cases to which we hope to apply the techniques. For these are cases where means are adjusted to ends and where unwanted side effects tend to be eliminated wherever possible, either by following an explicit plan or by less systematic fiddling. (1999, 118)

A similar argument is made by Kevin Hoover.

> Spirtes et al. (1993, 95) acknowledge the possibility that particular parameter values might result in violations of faithfulness, but they dismiss their importance as having "measure zero." But this will not do for macroeconomics. It fails to account for the fact that in macroeconomic and other control contexts, the policymaker aims to set parameter values in just such a way as to make this supposedly measure-zero situation occur. To the degree that policy is successful, such situations are common, not infinitely rare. (2001, 171)

Cartwright and Hoover both make it clear that they do not mean to say that the FC is *always* false, but only that it fails in certain situations, namely, those in which there is some process that selects for canceling out causal paths. For instance, in the exercise-weight example it seems unlikely that there is selection in favor of precisely counterbalancing parameterizations. Hence, their argument would not support the conclusion that exceptions to the FC are probable in that case. However, they do think that there is often selection for counterbalancing paths, which would imply that the FC is problematic as a general principle.[16]

If this objection is right, then there must be an assumption of SGS's theorem that is false in circumstances of the sort Cartwright and Hoover indicate. It is easy to see that the assumption called into question by Cartwright and Hoover's line of argument is **L**.[17] They claim that when there is selection for parameterizations in which causal paths cancel out, it is, for instance, probable in the linear model above that $b + ac = 0$. But the subset consisting solely of parameterizations that make $b + ac = 0$ is a two-dimensional subset of the three-dimensional parameter space. Hence, if it is probable that the actual parameterization is within that subset, then it is false that the probability of every subset of $n - 1$ dimensionality or less is zero.

Indeed, it would be unreasonable to maintain that $n - 1$ dimensional subsets of n-dimensional spaces must *always* have zero probability. For example, such a claim would entail that we must be certain a priori that no quantity is equal to any other quantity. This point can be appreciated by reference to the diagram in Figure 4.12. In the diagram, the subset of pairs of values in which a equals b is represented by the diagonal line in the square, which of course is one dimension less than the two-dimensional plane. Consequently, SGS's theorem can serve as a motivation for the FC only provided there is some explication of the conditions under which **L** is true and of why we should think that those conditions hold in the domain of application of the FC.[18] I suggest that **L** is a reasonable assumption when parameter values are affected by a large number of uncontrollable factors, a situation which is the norm in heterogeneous populations.

Consider a social planner attempting to do what Hoover describes, that is, to create a compensating mechanism to precisely counteract an

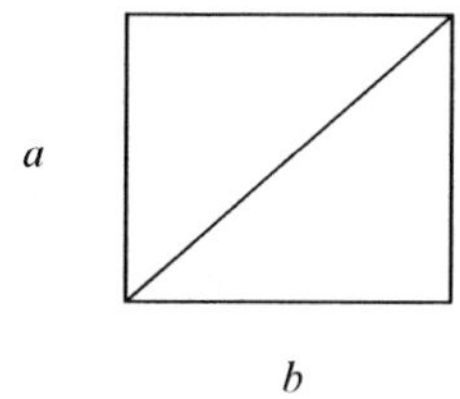

Figure 4.12 The Subset in which $a = b$

undesired side effect of some policy. For example, imagine a road improvement program that involves resurfacing and widening a number of large thoroughfares and some smaller side streets. Although improved, safer roads contribute to fewer traffic accidents, they also have the unfortunate side effect of increasing speeding, which is a significant cause of traffic fatalities. Letting R, S, and T be variables denoting road improvement, rates of speeding, and traffic fatalities, respectively, the example is represented by the graph in Figure 4.13. Suppose that, initially, the net effect of the road improvement is to increase the rate of traffic fatalities. To offset this problem, more police are hired to patrol the newly improved roads and the fines for speeding are increased. However, given a tight budgetary situation, the social planners do not want to spend more money on speeding prevention than necessary. They want to do just enough to make the two causal paths cancel out, and no more.

The strategy of the social planners in this case is to implement changes in the situation that will weaken the positive influence of R upon S so as to even the balance between the two paths. In principle, if the strength of influence of R upon S can be fine-tuned independently of the other parameters, this would be possible. But the relevant question is whether the social planners really can reliably make the exact canceling out occur, or at least be sufficiently approximated for practical purposes. Their ability to do so depends on being able to establish the following two things:

> *Selection of Parameters*: A process that tends to concentrate the weight of the distribution of parameterizations on a subset in which the FC is violated.

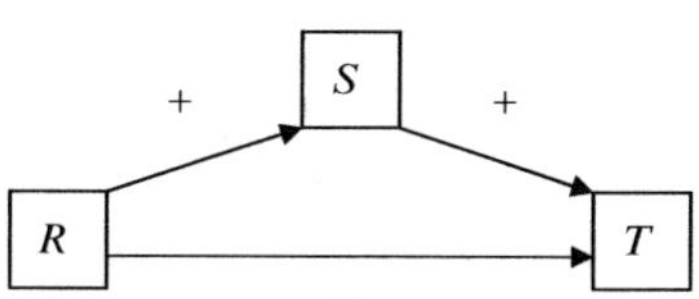

Figure 4.13 Road improvement and traffic fatalities

Homogeneity of Parameters: The absence of factors that perturb parameter values and thereby alter their distribution in uncontrolled ways.

If selection and homogeneity were perfectly accomplished, then the entire distribution of parameterizations would be restricted to an $n - 1$ dimensional subset of the parameter space. With regard to the graph in the linear causal model, an example of this would occur when it is certain that the parameterization makes $b + ac = 0$. Thus, when selection and homogeneity are perfectly satisfied, assumption **L** of SGS's theorem (that all $n - 1$ dimensional subsets of the parameter space receive probability zero) is false. Of course, it is unlikely that selection and homogeneity will be perfectly fulfilled in real-life examples. But if the distribution of parameterizations were tightly focused on an $n - 1$ dimensional subset of the parameter space rather than completely restricted to that subset, very near violations of the FC would be probable even if **L** were true. Moreover, there is little practical difference between *strict* and *very near* failures of the FC, since in either case, causal connections fail to give rise to correlations that are detectable in any obtainable sample size. Thus, Cartwright and Hoover's objection would be vindicated if very near exceptions to the FC were common in the principle's intended domain of application.

The above considerations show that selection and homogeneity suffice to make (near) exceptions to the FC probable. And Cartwright and Hoover's objection points out that it is not rare that someone or something endeavors to put a selection process in place. Trying and succeeding, however, are two very different things. In the road improvement example, it was assumed that the policymakers endeavored to make the causal paths cancel out through an adjustment of the strength of influence of R upon S. Bringing about the desired balance, then, requires knowing the requisite value of this parameter and being able to fine-tune it accordingly. Yet it is far from clear that this knowledge or ability is typically possessed by policymakers. How many additional police cruisers patrolling the streets would be required to reduce the value of the parameter by a given amount, for instance? For the moment, let us put aside this concern and suppose that the policymakers can devise a selection process.

Even if there is a process at work that tends to focus the probability distribution of parameterizations around an FC-violating subset, it does not follow that exceptions or near exceptions to the FC are probable, since the distribution of parameters might also be influenced by other trends that undo the work of the selection process. Suppose that there is a wide variety of difficult-to-predict or -control factors at play that are capable of altering the values of the parameters (i.e., that homogeneity does not hold even approximately). Clearly, these disturbing factors would be expected to increase the variance of the distribution of parameterizations, thus increasing the chance that the actual parameterization would fall in a region distant from a subset in which the FC is false. In addition to enlarging the variance of the distribution, factors that alter the values of parameters can also change its mean if not all parameters are uniformly susceptible to disturbance. For instance, if some parameters are more susceptible than others to factors that alter their values in a particular direction, then the mean of the distribution might be driven away from an FC-violating subset. In the road improvement example, if the effect of *R* upon *S* is sensitive to factors that tend to increase its value while the other parameters are relatively stable, then the mean of the parameterizations will move toward a positive net effect of *R* upon *T*.

The simple moral, then, is that the existence of a selection process can fail to make exceptions or near exceptions to the FC probable when a variety of uncontrollable factors that perturb parameter values is present.[19] Consequently, noting the presence of a selection process does not suffice to show that (near) violations of the FC are likely to occur. Yet Cartwright and Hoover's objection merely points out that it is common for selection processes to be present or at least for some effort to be made to create them, and then concludes that exceptions or near exceptions to the FC are likewise commonplace. Consequently, their argument is invalid on two grounds. First, effectively designing and implementing a selection process may be very difficult, so the fact that there is some effort afoot to create a selection process provides little assurance that one exists.[a] Second, even if a selection processes were common, Cartwright and Hoover's conclusion would follow only when homogeneity obtains. Yet it is obvious that the opposite is typically the case for the heterogeneous populations that are the concern of this book. Causal relationships in biological and social phenomena generally depend upon variable factors that are difficult to predict or control. Hence, the intended domain of the FC for the present purposes consists of causal systems of which it is quite doubtful that homogeneity is typically true or even approximately true.[20]

In short, Cartwright and Hoover's objection has failed to show that exceptions or near exceptions to the FC are common in its intended domain of use. Nevertheless, it would be a mistake to conclude that the FC is *always* an unproblematic assumption with regard to complex systems. In particular, near violations of the FC are probable when both selection and homogeneity are approximated, and it is arguably the case that this situation is not infrequent in gene knockout experiments.

Although it is not an example that they discuss, genetic redundancies illustrate the type of situation in which Cartwright and Hoover claim that exceptions to the FC are probable. For example, imagine a gene that serves as a template for the transcription of a protein that normally performs a specific set of functions in a cell, but when that protein is not present in sufficient quantities, the transcription of a distinct yet functionally similar protein from a second gene is increased. Moreover, it is plausible that there would be an adaptive benefit in the quantitative strengths of the two paths exactly counterbalancing one another. For example, maintaining the function may require that the sum of two products be kept within fairly narrow bounds. Hence, it would not be optimal for both genes to normally be transcribed together, while it would be beneficial that the function be maintained at the normal rate when the usual product is not present in adequate quantities. Thus, natural selection would constitute a process that favors parameterizations in which the counteracting paths exactly or very nearly cancel out.[21]

Moreover, apparent near exceptions to the FC are not rare in gene knockout experiments. For instance, a recent gene knockout study (Scarff et al. 2004) examined a particular protease inhibitor, SPI3, believed to have several important functions which primarily involve preventing certain proteases from affecting nontarget cell and tissue types. The investigators produced a strain of mice in which the gene from which SPI3 is transcribed was disabled, but surprisingly this mutant strain appeared completely normal and showed no apparent difference in any of the several functions to which SPI3 is believed to be relevant. Given the FC, this result would constitute strong evidence that SPI3 is not a cause of any of the functions in question. However, that was not the conclusion drawn by the researchers. They noted that among mice in which the SPI3 gene had been knocked out, the presence of a second protease inhibitor, EIA, was increased. Since EIA is functionally similar to SPI3, this suggested that the failure of the gene knockout to produce any detectable difference between the mutant and wild-type strains could be explained by a compensating pathway.

The authors suggested two possible mechanisms through which the knockout of the gene for SPI3 could stimulate the increased transcription of EIA (ibid., 4080). In both cases, higher levels of SPI3 inhibit the transcription of the gene from which EIA is synthesized, thereby suppressing EIA under normal circumstances. The basics of the hypothesis, then, can be represented in the graph in Figure 4.14. The variables G_{SPI3} and G_{EIA} represent the rate of transcription of the genes for SPI3 and EIA, respectively. According to the authors, the results of their experiment "indicate that EIA levels are increased in SPI3-deficient mice to compensate for the loss of SPI3" (ibid., 4079).

Nor is the above example an aberration.[22] As Sandra Mitchell (2003, 154–55) notes, redundancy is a common challenge for gene knockout

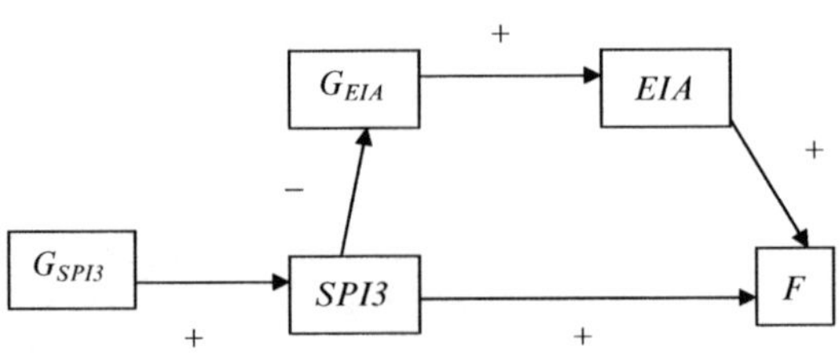

Figure 4.14 An example of genetic redundancy

experiments. Indeed, a recent issue of *Nature* included a news feature on exactly this topic. It pointed out that:

> In many cases, a mutant mouse [produced by the gene knockout] does not show any obvious characteristics—or phenotype. In others, the phenotype disappears when the disabled gene is crossed into a different strain of mouse. Indeed, clear and consistent phenotypes now seem to be the exception rather than the rule. (Pearson 2002, 8)

A common explanation given for this state of affairs is genetic redundancy: "These results often reflect the fact that genes acting in parallel pathways can compensate for the one that is missing" (ibid.).[23] Moreover, the journal *Molecular and Cellular Biology* has, since November 1999, dedicated a section of each issue to the topic of gene knockout studies that find surprising differences, or lack of differences, between mutants and controls.[24] That was where the study described above was published.

As explained above, the presence of a selection process alone is not sufficient to make exceptions or near exceptions to the FC probable; homogeneity is also required. But this latter condition is much more likely to be approximated in the context of a gene knockout experiment than in, say, a wild population of mice. The mice in knockout experiments are typically generated from extremely genetically homogeneous strains that have been reared for numerous generations under standard laboratory conditions. The procedure by which knockout strains are generated further enforces this homogeneity. Knockout strains of mice are generated by disabling the target gene in embryonic stem (ES) cells and then either injecting these ES cells into a blastocyst or aggregating them with an embryo at an earlier stage.[25] Since they are the product of cells from more than one individual, such modified embryos are known as chimeras. Chimeras formed by blastocyst injection or aggregation at earlier stages will, if viable, transmit the modified genes through the germ line. The knockout strain can then be generated from the mutant germ line cells selected from the chimeras (if these can be successfully created). The knockout strains, therefore, are generated not only from strains that are quite genetically homogeneous, but also in a way that produces an effective genetic bottleneck, since the knockout strain ultimately derives from the chimeras, which descend from the modified ES cells and their zygote hosts.

Gene knockout experiments, then, are a context in which it is not unlikely that both selection and homogeneity are satisfied to a reasonably good approximation, and hence near exceptions to the FC are probable. This analysis of failures to find effects in gene knockout experiments has an empirical consequence. In cases in which there is good reason to believe that a causal mechanism is present, despite the null result of a knockout experiment, it is likely that a probabilistic dependence between the suspected cause and effect will appear if the conditions of the experiment are varied. And in fact, surprising absences of difference between mutants and controls in one gene knockout experiment often emerge in other mouse strains, or strains reared in distinct environmental conditions (Pearson 2002, 8–9).

That the FC is a problematic assumption in the case of gene knockout experiments does not show that it is an inappropriate assumption generally. For example, the challenge to the FC arising in gene knockout experiments should not be expected to transfer to studies involving more heterogeneous populations, such as human subjects, for the reasons discussed above. Thus, the lack of genetic and environmental homogeneity among experimental subjects, surprisingly enough, can facilitate the discovery of causal structure in virtue of providing a more hospitable setting for the FC. Of course, this is not to deny that there are genuine benefits of uniform populations of experimental subjects, but it does show that experiments involving such subjects also have some potential downsides. It is not too difficult to see how those downsides can be avoided in the present case: vary the genetic and environmental backgrounds of the experimental populations. But although this prescription is simple enough in principle, there are practical obstacles to implementing it in the laboratory. As the news feature from *Nature* cited above observes:

> Ideally, experiments on knockout mice would routinely include work on multiple strains. In practice, most researchers in the field ague that this is not realistic—creating a single knockout strain can take up the majority of a three-year PhD project. (Pearson 2002, 8)

Barring the development of methods that allow knockout strains to be created more easily, the FC seems likely to remain a problematic assumption in gene knockout experiments for the foreseeable future.

So, where does all this leave the disruption principle? Recall that the disruption principle has two parts. First, if there is no undisrupted mechanism from *X* to *Y*, then ideal interventions on *X* do not alter the probability distribution of *Y*. As explained in section 4.4.1, this part rests on solid ground. Second, the disruption principle asserts that if there is an undisrupted mechanism from *X* to *Y*, then ideal interventions on *X* alter the probability distribution of *Y*. In this section I have argued that, some objections notwithstanding, this is normally an appropriate assumption with respect to complex causal systems wherein the strengths of causal influences are subject to a wide array of uncontrolled factors. Nevertheless, there are some situations, as illustrated by gene knockout experiments, in which this heterogeneity is substantially reduced, and hence in which near exceptions to the disruption principle are more probable.

4.5 CONCLUSION

This chapter introduced, illustrated, and explored the range of applicability of the disruption principle, a central premise of the mechanisms approach to extrapolation. It was shown how this principle could be stated by way of the formalism of directed graphs, and its role was illustrated by reference to an example drawn from HIV research. It was also shown that the principle is a logical consequence of two other, more familiar propositions connecting causality and probability: the Principle of the Common Cause (PCC) and the Faithfulness Condition (FC). The aspect of the PCC specifically relevant to the disruption principle was shown to be unproblematic, but the case of the FC was more complex. Although there is a good motivation for the FC for heterogeneous, naturally occurring biological populations, that justification does not extend to extremely homogeneous populations, such as the strains of laboratory mice typically used in gene knockout experiments. This result delineates more exactly the conditions under which the FC is and is not an appropriate methodological principle of causal inference, and it shows that the experimental practice of holding all background factors fixed is not always a virtue in the study of complex systems, since such experimental arrangements may conceal probabilistic dependencies between cause and effect that are present in messier, real-world populations.

5

Extrapolation, Capacities, and Mechanisms

Imagine that a chemical occurring in some consumer products has been found to be carcinogenic if administered in large doses in rats, and the question is whether it is also a carcinogen in humans. The mere repetition of the experimental result in rats is not sufficient to answer this question, since the physiology of rats may differ in some relevant respect from that of humans. This is an example of extrapolation: given some knowledge of the causal relationship between *X* and *Y* in a base population, we want to infer something about the causal effect of *X* upon *Y* in a target population.[1] For instance, in the example above, we know that the chemical is a positive causal factor for cancer in rats and we want to know whether it is also such in humans. Difficult cases of extrapolation are ones in which the base and target populations may differ in relevant respects and, moreover, in which ethical or practical considerations prohibit directly testing the claim at issue by experiment in the human target population.

The most straightforward way to approach extrapolation is to presume that what is true of one population is also approximately true of other related populations unless there is some specific reason to think otherwise. I call this inferential strategy *simple induction*. However, since simple induction would inevitably lead to many mistaken extrapolations, a more sophisticated approach would be highly desirable. Any account of extrapolation that goes beyond simple induction must confront two basic challenges. The first is what I call the *extrapolator's circle*. Simple induction relies on some criterion of relatedness, such as phylogeny or type of economic system. The shortcomings of simple induction stem from the fact that satisfying such criteria is often not sufficient for being a reliable basis for extrapolation. Consequently, additional information about the similarity between the model and the target—for instance, that the relevant mechanisms are the same in both—is needed to justify the extrapolation. The extrapolator's circle is the challenge of explaining how we could acquire this additional information, given the limitations on what we can know about the target. In other words, it needs to be explained how we could know that the model and the target are similar in causally relevant respects without already knowing the causal relationship in the target. The second challenge arises from the inevitable presence, in the biological and social sciences, of causally relevant differences between the model and the target. Thus, any adequate account of extrapolation in heterogeneous populations must explain how extrapolation can be possible even when such differences are present.

I argue that existing accounts of extrapolation fail to answer these challenges. One proposal I consider maintains that capacities or causal powers that exert a characteristic influence independently of context are a basis for extrapolation. However, this proposal does not adequately explain how one is to know that one is dealing with a capacity rather than a context-sensitive causal relationship, aside from already having found that the causal relationship obtains in all of the contexts in question. Thus, without some further elaboration, the capacities proposal does not address the two challenges just described. A mechanisms approach to extrapolation could be regarded as such an elaboration of the capacities proposal or as a separate approach. According to this approach, knowledge of mechanisms linking cause and effect and knowledge of factors capable of interfering with these mechanisms can significantly facilitate extrapolation. As noted in Chapter 1, this idea has been suggested by several philosophers, social scientists, and biologists. However, a mere invocation of mechanisms does not resolve the extrapolator's circle, nor does it explain how extrapolation can be possible in the face of causally relevant disanalogies. The mechanism approach needs to explain how the suitability of the model could be established without already knowing all of the important details about the mechanism in the target. Moreover, some differences in the mechanism in the model and the target are inevitable in biology and social science. Thus, the mechanisms approach requires an account of how extrapolation can be justified even when such differences are present.

In this chapter, I develop a more satisfactory version of the mechanisms approach to extrapolation. The central concept is a mode of inference I call *comparative process tracing*, which aims to assess the suitability of the model as a basis for extrapolation. Comparative process tracing depends upon background information concerning likely similarities and differences between the model and the target. If significant differences between the model and the target are likely to be restricted to a relatively small number of stages of the mechanism, then comparisons at those stages may provide good grounds for the suitability of the model. The number of stages that must be compared can be reduced further if upstream differences must result in differences at an observable downstream point in the mechanism. Moreover, knowledge of just a few stages of the mechanism in the target al*one* (that is, without knowledge of the model) might not suffice for firm conclusions regarding the existence of a mechanism in the target. Hence, my proposal provides an analysis of how extrapolation can be justified despite the extrapolator's circle while indicating conditions in which it is a genuine problem. I illustrate my account of comparative process tracing with a case study concerning the carcinogenic effects of aflatoxin B_1. The question of how useful the mechanisms approach to extrapolation developed here is likely to be in social science is taken up in Chapter 8.

The aflatoxin example also illustrates how extrapolation may be justifiable even when there are some causally relevant disanalogies between the model and the target. I argue that the closeness of the match between model and target required for extrapolation depends upon the specificity of the causal claim to be extrapolated. While similarity in all causally relevant respects may be required for extrapolating an exact, quantitative causal effect, it is not required for extrapolating qualitative causal claims. In particular, claims about positive or negative causal relevance can be extrapolated even when there are causally relevant disanalogies. This point is illustrated by the aflatoxin example, wherein a causally relevant difference between the animal model and the human suggests that the carcinogenic effect is less in the model than in humans. Yet this difference does not indicate that it would be a mistake to extrapolate the claim that exposure to aflatoxin B_1 increases the chance of liver cancer in humans. A more general and precise characterization of conditions that suffice for extrapolating claims about positive and negative causal relevance is given in Chapter 6.

5.1 SIMPLE INDUCTION

Imagine a case in which one is concerned to decide whether a causal generalization found in a base population (say, laboratory mice) also holds true of a target population of interest (say, humans). Simple induction proposes the following rule for such cases:

> Assume that the causal generalization true of the base population also holds approximately in related populations, unless there is some specific reason to think otherwise.

In other words, simple induction proposes that extrapolation be treated as a default inference among populations that are related in some appropriate sense. The advantage of simple induction is that it can be employed in cases in which relatively little detailed information concerning the mechanisms underlying the causal relationship is available. There are, however, three aspects of the above characterization of simple induction that stand in obvious need of further clarification. In particular, to apply the above rule in any concrete case, one needs to decide what it is for a causal generalization to hold approximately, to distinguish related from unrelated populations, and to know what counts as a reason to think that the extrapolation would not be appropriate. It seems doubtful that a great deal can be said about these three issues in the abstract—the indicators of related populations, for instance, can be expected to be rather domain-specific. But it is possible to give examples of the sorts of considerations that may come into play.

Simple induction does not enjoin one to infer that a causal relationship in one population is a precise guide to that in another—it only licenses the conclusion that the relationship in the related target population is "approximately" the same as that in the base population. It is easy to see that some qualification of this sort is needed if simple induction is to be reasonable. In biology and social science, it is rare that a causal effect in one population is *exactly* replicated even in very closely related populations, since the probabilities in question are sensitive to changes in background conditions. Nevertheless, it is not rare that various qualitative features of a causal effect, such as positive relevance, are shared across a wide range of populations. For example, tobacco smoke is a carcinogen among many human and nonhuman mammal populations. Other qualitative features of a causal effect may also be widely shared; for instance, a fertilizer may promote growth in moderate dosages and inhibit growth in large ones across a wide variety of plant species even though the precise effect differs from one species and variety to the next. In other cases, the approximate similarity may also refer to quantitative features of the causal effect—the quantitative increase in the chance of lung cancer resulting from smoking in one population may be a reasonably good indicator of that in other closely related populations. In the case of extrapolation from animal models, it is common to take into account scaling effects due to differences in body size, since one would expect that a larger dose would be required to achieve the same effect in a larger organism (cf. Watanabe et al. 1992). Thus, in such cases, the scaling adjustment would constitute part of what is covered by "approximately." Depending on the context, the term "approximate" could refer to similarity with regard to any one of the aspects of the causal effect mentioned above, or other aspects, or any combination of them.

Simple induction is also restricted in allowing extrapolations only among related populations, a qualification without which the rule would obviously be unreasonable: no population can serve as a guide for every other. In biology, phylogenetic relationships are often used as a guide to relatedness for purposes of extrapolation: the more recent a shared common ancestor, the more closely related the two species are (cf. Calabrese 1991, 203–4). A phylogenetic standard of relatedness also suggests some examples of what might count as a specific reason to think that the base population is not a reliable guide for the target population. From the mechanistic point of view, phylogenetic relatedness supports extrapolation because it increases the likelihood that the pertinent mechanisms are shared in the base and target populations as the result of descent from a common ancestor. But when the causal relationship in the base population depends on derived features—that is, characteristics not inherited from the common ancestor—this reasoning is fallacious.

In many biological examples, simple induction requires only some relatively minimal background knowledge concerning the phylogenetic relationships among the base and target populations, and its chief advantage lies in this frugality of information demanded for extrapolation. Yet the weakness of the simple inductive strategy also lies in exactly this frugality: given the rough criteria of relatedness, the strategy will inevitably produce many mistaken extrapolations. According to one review of results concerning interspecies comparisons of carcinogenic effects:

> Based on the experimental evidence from the CPDB [Carcinogenic Potency Database] involving prediction from rats to mice, from mice to rats, from rats or mice to hamsters, and from humans to rats and humans to mice,...one cannot assume that if a chemical induces tumors at a given site in one species it will also be positive and induce tumors at the same site in a second species; the likelihood is at most 49%. (Gold et al. 1992, 583)

A related challenge for simple induction is that it is not rare that there are significant differences across distinct model organisms. For instance, aflatoxin B_1 (discussed in section 5.3.2) causes liver cancer in rats but has little carcinogenic effect in mice (Gold et al. 1992, 581–82; Hengstler et al. 2003, 491).

The consequence of these considerations is not that simple induction is wrong or useless for extrapolation. Rather, what follows is that simple induction is limited, and that it is highly desirable that it be supplemented with some more sophisticated inferential strategy. Let us turn to the question of just what this "something more" should be.

5.2 POWERS AND CAPACITIES

The notion of a causal power or capacity is a very commonsensical one. For example, in virtue of its hardness and mass, a brick has the capacity to shatter a glass window. Moreover, this capacity is not tied to a specific set of background conditions, but is something that the brick can be reasonably be expected to possess in whatever circumstance it is likely to be found. Capacities and causal powers, then, seem like a promising point of departure from which to address extrapolation. For example, Cartwright maintains that it is only knowledge of capacities that enables one to extrapolate context-dependent relationships such as causal effects from one population to another (1989, 157–58, 163; 1992, 56). According to Cartwright, a statement about a capacity tells us what would occur when all other causes are absent (cf. 1992, 49; 1999, 82–83). But it tells us more than just that, since a capacity exerts its characteristic influence upon the effect even when other causes are present (ibid.). In this section, I argue that capacities approaches to extrapolation have failed to overcome the limitations of simple induction.

The central feature of capacities is their stability across changes in background conditions. As Cartwright puts it, "A property carries its capacities with it, from situation to situation" (1989, 146). This stability need not be absolute (cf. Cartwright 1989, 163), but it is presumably required to be sufficiently robust to justify the expectation that causal influence will hold throughout the domain in question. Capacities are not limited to basic physical properties, such as the mass of a brick. Cartwright also uses the term "capacity" to refer to causal relationships that depend on a complex set of interactions. For instance, "aspirin's capacity to relieve headaches" is one of her stock illustrations (1989, 141). A claim about the palliative effects of aspirin is quite similar to a statement about the carcinogenic effects of a particular chemical compound. Both are claims about positive causal relevance that depend upon an interaction between a compound and an organism. Thus, the palliative virtues of aspirin exist only in relation to organisms with a particular type of physiology, and similarly for the carcinogenic effects of a particular compound.

Clearly, it will often be difficult to know in advance whether a compound that has a particular effect in one species or class of organisms will have a similar effect in others. That of course is the extrapolation problem of concern in this chapter. But by definition, a capacity is a causal influence that is not tied to a specific context. Hence, if we know *only* that the compound is carcinogenic in (say) rats, we do not know whether its influence can be properly called a capacity. Consequently, if we do not know whether the extrapolation would be correct, we do not know whether the causal effect in question is a capacity. The difficulty here, then, it is that it is not clear how one is to know that something is a capacity independently of already knowing what one wanted to know about extrapolation. In other words, to call a causal relationship a capacity is to say that it is stable across a range of contexts of interest, but questions of extrapolation arise exactly in those cases in which the stability of the causal relationship is in doubt.[2]

The objection that it is unclear how one is supposed to know whether a causal relationship is a stable capacity or merely a local, context-dependent effect has been raised by several authors (Morrison 1995, 165–66; Glennan 1997, 611–13). In response to such concerns, Cartwright writes:

> I have claimed that in the central uses of the concept, we assume that within the specified domain tendencies when properly triggered always "contribute" their characteristic behaviours unless there is a reason why not. (1995, 180)

This statement amounts to a commitment to the use of simple induction: within some set of related populations (the domain), one assumes that the relationship holds unless there is some reason to suppose otherwise. However, we have seen that simple induction is often highly problematic in the context of extrapolation from animal models. Hence, without some further elaboration, the capacities approach will not suffice as a normative account of extrapolation.

Cartwright does provide some elaboration on the issue of whether there is a reason why the capacity will not operate in the new context. This judgment is said to be based upon knowledge of "how this tendency naturally operates and how its power to do so is transmitted, what could distort it, what enhance it, what could damp it and in what ways" (ibid.). This appears to be a reference to Cartwright's notion of a nomological machine, which is one of several related mechanism concepts, as was discussed in section 3.4.1. On this proposal, then, capacities inhere in the component parts of a nomological machine or mechanism, while extrapolation depends upon information about how the component parts are arranged and interact. That is very similar to the suggestion that knowledge of mechanisms and interfering factors is a basis for extrapolation. But merely to invoke mechanisms is not to have explained how the challenges confront extrapolation. For all we know, the causal effects of the components of the mechanism might be context-dependent, and the components in the model might be arranged and interact differently than those in the target (Alexandrova 2006, 186–87). Demonstrating the relevant similarity of the model and the target would presumably require separately studying the mechanisms in both and then comparing results. But it is not clear how that can be done when the ability to study the target directly is severely limited. In short, to gesture toward mechanisms is not to have answered the challenges confronting extrapolation.

Cartwright's proposal is not unique in this regard: the same point can be made in the context of an account of causal powers provided by Patricia Cheng (1997). Although the aim of Cheng's approach is primarily the psychological one of understanding how people actually draw causal inferences, her proposal is highly interesting from a philosophical perspective. Like Cartwright, Cheng stresses the value of causal powers with regard to extrapolating causal conclusions.

> In the reasoner's mind, causal powers are invariant properties of relations that allow the prediction of the consequences of actions regardless of the causes of an effect (those other than the candidate causes) that happen to occur in a situation. (2000, 127)

Thus, Cheng's causal powers are very similar to Cartwright's capacities in that they are intended to be stable influences that operate independently of changes in context or background conditions. In its simplest version, Cheng's proposal assumes the existence of two types of causes, generative and preventive, which may be either present or absent (but not vary otherwise). No event occurs unless it is caused, and causes can influence their effects only when they are present. An event occurs if and only if at least one of its potential causes is present and causes it on that occasion. For a generative cause C, the causal power of C with respect to E, which we may denote by p_{ce}, is the probability that C causes E provided that C occurs. It need not be the case that $p_{ce} = P(E|C)$, since $P(E|C)$ depends not only upon the efficacy of C but also upon the probability of the presence and effectiveness of other causes of E.

Cheng's innovation is to demonstrate that, given certain assumptions, causal powers can be estimated from statistical data (1997, 373–74). In her 1997 paper, one of these assumptions is that p_{ce} is independent of the occurrence of all other causes of E. This means that C, in affecting E, does not interact with any other causes. But in biology and social science, causes typically influence their effects interactively, so that the impact of one depends upon the presence or absence of others. Indeed, extrapolation is difficult precisely because the relationship between cause and effect might depend on some unknown, variable factor. In light of this limitation, Cheng (2000) develops a concept of interactive causal power. She points out that when causes interact, the formula described in her original proposal does not estimate causal powers, but only what she terms the "contextual causal power" (2000, 235). Cheng also specifies conditions in which interactive causal powers can be estimated from statistical data, provided that *all* the interacting causes have been measured (2000, 241–46). However, in most interesting biological and social science examples, it can be expected that the causes under investigation interact with other causes that have not been measured or otherwise explicitly taken into account. For such cases, Cheng suggests that one proceed by first assuming that the causal power is simple (that is, - independent of all other causes) and then postulate causal interactions only when necessary to accommodate conflicting data (2000, 232, 238). Yet the proposal that one estimate context- or population-sensitive causal relationships, and then assume that these hold approximately in related populations unless there is some evidence to the contrary, is simple induction. And as explained in the foregoing section, simple induction is often not a sufficient basis for extrapolation from animal models. Thus, the proposals considered in this section have not provided an adequate account of how extrapolation could proceed even when not justifiable by simple induction. Let us turn, then, to a distinct proposal.

5.3 MECHANISMS-BASED EXTRAPOLATION

The mechanisms approach to extrapolation suggests that knowledge of mechanisms and factors capable of interfering with them can provide a basis for extrapolation. But this proposal must also answer the two challenges to extrapolation described above. Since causally relevant differences between model and target are inevitable, some explanation must be provided of how extrapolation can be justified even when there are some differences in mechanism between model and target. The extrapolator's circle confronts the mechanisms proposal as well. Presumably, justifying the appropriateness of the model would involve comparing mechanisms in the model and the target, which would involve independently studying the mechanisms in both and then comparing results. But that makes it unclear how the suitability of the model can be established without already knowing what the extrapolation was supposed to tell us. In this section, I argue that existing discussions of mechanisms do not adequately address these challenges. Then I present a more adequate account of mechanisms-based extrapolation that is founded upon what we call *comparative process tracing*.

5.3.1 The Existing Literature on Mechanisms and Extrapolation

There is a small literature that provides detailed case studies of extrapolation in biology or social science, often with particular attention to the role of mechanisms (cf. Burian 1993; Ankeny 2001; Schaffner 2001; Weber 2005; Guala 2005; Alexandrova 2006). Essays in this genre point out some circumstances that facilitate, and some that hinder, extrapolation. For instance, it has been observed that extrapolation is on firmer ground with respect to basic, highly conserved biological mechanisms (Wimsatt 1998; Schaffner 2001; Weber 2005, 180–84). Others have observed that a close phylogenetic relationship is not necessary for extrapolation and that the use of a particular animal model for extrapolation must be supported by empirical evidence (Burian 1993). Similarly, Francesco Guala (2005) emphasizes the importance in experimental economics of providing empirical evidence to support the claim that the model is relevantly similar to the target.

These suggestions are quite sensible. The belief that some fundamental biological mechanisms are very widely conserved is no doubt a motivating premise underlying work on such simple model organisms as the nematode worm. And it is certainly correct that the appropriateness of a model organism for its intended purpose is not something that may merely be assumed, but a claim that requires empirical support. Yet such observations do not answer the challenges to extrapolation. Objections to animal extrapolation focus on causal processes that do not fall into the category of fundamental, conserved biological mechanisms. For example, Marcel Weber suggests that mechanisms be conceived of as embodying a hierarchical structure, wherein the components of a higher-level mechanism consist of lower-level mechanisms, and that while lower-level mechanisms are often highly conserved, the same is not true of the higher-level mechanisms formed from them (2001, 242–43; 2005, 184–86). So, even if one agreed that basic mechanisms are highly conserved, this would do little to justify extrapolations from mice, rats, and monkeys to humans regarding such matters as the safety of a new drug or the effectiveness of a vaccine. Since critiques of animal extrapolation are often motivated by ethical concerns about experimentation on animals capable of suffering (cf. LaFollette and Shanks 1996), they primarily concern animal research regarding less fundamental mechanisms that cannot be studied in simpler organisms such as nematode worms or slime molds. Nor do the observations sketched in the foregoing paragraph explain how extrapolation can proceed even when there are causally relevant differences between model and target or how the extrapolator's circle is to be avoided. For example, noting that the appropriateness of an animal model for a particular extrapolation is an empirical hypothesis does not explain how such a hypothesis can be established without already knowing what one wishes to extrapolate.

There also are discussions in the philosophical literature of strategies for learning about mechanisms. A distinction between mechanisms and the "phenomena" (Craver and Darden 2001, 113–14) or "behavioral descriptions" (Glennan 2005, 446) those mechanisms explain is helpful for understanding these proposals and contrasting them with comparative process tracing. Phenomena are regularities of the system under study that are more easily observable than the underlying mechanisms. For example, that HIV exposure causes AIDS is a phenomenon, whereas the mechanism consists of the molecular processes through which HIV has this effect. Since phenomena are often more easily discovered than underlying mechanisms, several authors have examined strategies for discovering mechanisms, given the phenomenon and some background constraints on what the components of the mechanism and their interactions could be (cf. Bechtel and Richardson 1993; Craver and Darden 2001; Darden and Craver 2002). Lindley Darden and Carl Craver's (2002) discussion focuses on what they term *schema instantiation* and *forward chaining/backtracking*. Schema instantiation begins with a schematic outline of the mechanism in which central functional roles are specified, but important details concerning the entities and activities involved in the performance of those functions are omitted. For example, the mechanism of HIV replication instantiates a schema that is common for retroviruses: attachment to a target cell, insertion of viral RNA into cytoplasm, reverse transcription, integration of viral DNA into host DNA, and synthesis of products for the formation of new viruses from this integrated viral DNA by means of the host cell's genetic machinery. Next, one attempts to discover the specific entities and activities that instantiate the schema, often by means of tracing forward from a known starting point or backward from a known end point (or both at once). For convenience, we will refer to the joint application of these strategies, schema instantiation and forward chaining/backtracking, as *process tracing*.

Glennan points out that process tracing is sometimes unfeasible for ethical or practical reasons (2005, 459–61). In such cases, one may attempt to discover the mechanism through more detailed descriptions of the phenomenon (ibid.). For example, alternative hypotheses concerning the mechanism may yield differing predictions about how the system would behave in a new circumstance. But although the strategy that Glennan suggests differs from process tracing, it is aimed at solving the same inference problem: *given* a description of the phenomenon, *discover* the mechanism that accounts for it. In extrapolation, by contrast, what one wishes to infer is a mechanism and phenomenon in a target organism. The evidence given includes the mechanism and behavioral description for a model organism, and perhaps some partial information about the mechanism in the target. By "partial information," I mean that the information concerning the mechanism in the target is not sufficient on its own to infer the phenomenon (e.g., whether the compound is carcinogenic in humans). The mechanisms approach to extrapolation must indicate a strategy for solving the following inference problem: *given* both the mechanism and the phenomenon *in the model*, and partial information concerning the mechanism in the target, *infer* the mechanism and/or phenomenon *in the target*.

5.3.2 Comparative Process Tracing

Suppose that one is given a description of the mechanism in the model organism and wishes to use this information as a basis for extrapolation. Such an inference is a case of reasoning by analogy. The form of arguments by analogy can be represented schematically as follows: the base (or source or analogue) is known to possess properties 1 through n, while the target is known to have properties 1 through $n-1$; therefore, the target also possesses property n. It is obvious that not all inferences satisfying this abstract schema are reliable. For instance, Bob and Sue may both own 2005 Volkswagen Beetles, yet the information that Bob's car is iridescent lime green provides little support for the conclusion that Sue's car is the same color (cf. Weitzenfeld 1984, 138; Davies 1988, 229). Arguments instantiating the above schema, then, provide substantial support for their conclusions only given some additional, perhaps implicit, information. This additional information would consist of generalizations asserting that objects of specified types typically resemble one another in certain ways, though not necessarily in others. For instance, suppose one wanted to know whether the engine in Sue's Volkswagen Beetle is in the rear of the car (as in the older models) or in the front. If we learned that the engine of Bob's car is front mounted, we readily conclude that the same is true of Sue's car. The difference between this analogical inference and the one above is that cars of the same make, model, and year are typically manufactured in a variety of colors yet are generally similar with regard to basic design features such as the placement of the engine. Likewise, mechanisms-based extrapolation depends on knowledge of likely similarities and dissimilarities of the mechanisms between model and target.

If one peruses a text or review article on animal extrapolation in toxicology, one finds a compendium of information concerning how pertinent mechanisms differ between humans and various model organisms, and with respect to which types of compounds.[3] In the case of carcinogenesis, probably the most frequent differences concern metabolism (Calabrese 1991, chap. 5; Hengstler et al. 1999, 918). Since the metabolism of foreign, potentially toxic compounds consists of chemically transforming them so as to make them less toxic and more readily excreted, differences with regard to how a particular compound is metabolized, and at what rate, can have implications for its carcinogenic effects. Mechanisms for metabolism of foreign compounds are typically described in terms of two phases (cf. Calabrese 1991, 206). In phase I, the compound is chemically altered (often through the addition of oxygen or hydrogen atoms) in a manner that makes it more polarized, and consequently more easily excreted. In phase II, the compound resulting from the modification in phase I is conjoined with a macromolecule, such as a carbohydrate, which typically detoxifies the compound and further facilitates its removal. Metabolic mechanisms can differ with respect to how the compound is altered at either phase and in virtue of which enzymes catalyze the process, which has the result that some mechanisms may be more effective than others at detoxifying and eliminating a given foreign compound.

The above discussion suggests a procedure for extrapolating a mechanism found in the base population to the target population, a procedure that I call *comparative process tracing*. First, learn the mechanism in the model organism, by means of process tracing or other experimental means. For example, a description of a carcinogenic mechanism would indicate such things as the product of the phase I metabolism and the enzymes involved; whether the metabolite is a mutagen, an indication of how it alters DNA; and so on. Second, compare stages of the mechanism in the model organism with that of the target organism in which the two are most likely to differ significantly. For example, one would want to know whether the chemical is metabolized by the same enzymes in the two species, and whether the same metabolite results, and so forth. In general, the greater the similarity of configuration and behavior of entities involved in the mechanism at these key stages, the stronger the basis for the extrapolation.

The reliability of comparative process tracing depends on correctly identifying the points at which significant differences between the model and the target are likely to arise. Significant differences are those that would make a difference to whether the causal generalization to be extrapolated is true in the target. For instance, metabolism is a source of potentially significant difference in carcinogenesis, since how a compound is metabolized often matters to whether it is carcinogenic or not. Judgments about where significant differences are and are not likely to occur are based on inductive inferences concerning known similarities and differences in related mechanisms in a class of organisms, and on the impact those differences make. In the present case, the relevant generalizations would concern the common similarities and significant differences in carcinogenic mechanisms between humans and rodents. Comparative process tracing, then, resembles simple induction in relying upon generalizations concerning the relation between the target and model organisms. The chief difference concerns what these generalizations assert. Simple induction depends upon generalizations of the form "What is carcinogenic for rats is probably carcinogenic for humans, too." In contrast, comparative process tracing depends upon generalizations like "Features *A*, *B*, and *C* of carcinogenic mechanisms in rodents usually resemble those in humans, while features *X*, *Y*, and *Z* often differ significantly." The toxicology literature described above is plausibly interpreted as an effort to provide an empirical basis for generalizations of the latter but *not* the former sort. Of course, it might be questioned whether the data presently available to toxicologists constitute a representative sample. However, that is a standard problem of statistical sampling rather than

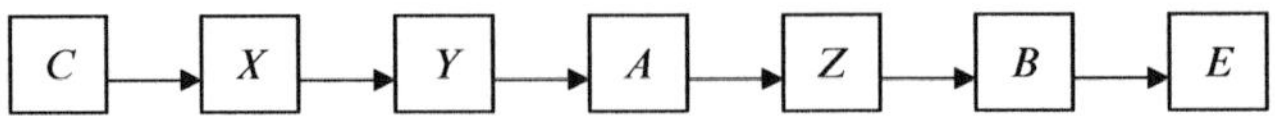

Figure 5.1 Comparing a downstream stage

a difficulty specifically raised by extrapolation, such as the extrapolator's circle.

But even given accurate information about the points of likely similarity and dissimilarity, comparative process tracing might still be impractical if not all likely points of significant difference could be compared. Fortunately, comparative process tracing often does not require comparing every stage of the mechanism at which significant differences are likely to be present. In particular, suppose that many points of likely difference are upstream of a later stage that is relatively easy to measure and compare. Then it may be possible to omit comparisons of the upstream stages and focus on the downstream one. For instance, imagine a mechanism like the following:

Suppose that *X*, *Y*, and *Z* represent points of the mechanism at which significant differences between model and target are likely, while *A* and *B* represent points that are likely to be the same. If differences in *X* or *Y* must result in differences in *Z*, then it is necessary only to compare the model and target at *Z*. That reduces the amount of information about the mechanism in the target that is needed to establish the suitability of the model, which may be very helpful if it is difficult to study the mechanism in the target directly. Furthermore, comparing a downstream stage of the mechanism also renders mistakes about upstream sources of difference less consequential. For instance, suppose that differences were in fact likely at *A* in Figure 5.1, despite our belief to the contrary. Yet if a difference at *A* must generate differences at *Z*, then the mistaken belief about *A* will not lead to a faulty extrapolation so long as a comparison is made at *Z*. Thus, efficient applications of comparative process tracing can focus on likely sources of difference in *downstream* stages of the mechanism.

A few important qualifications about the emphasis on downstream stages should be noted. First, the strategy could lead to mistaken conclusions if there is a path that bypasses the downstream stage. For instance, suppose in Figure 5.1 there was a path from *X* to *E* that did not go through *Z*. In that case, checking *Z* would not be sufficient since there might be significant differences in the mechanisms that would not leave a mark on *Z*. Hence, applications of the strategy depend on knowing where to look for bottlenecks through which any influence upon the outcome must be transmitted. Second, the mark that upstream stages leave upon the downstream stages must be distinctive in the sense that it could not have resulted from some independent cause. The mark should be, as it were, a fingerprint whose presence or absence indicates something causally significant about upstream processes. In examples from toxicology, the distinctive mark is often a particular chemical compound that retains a distinctive, identifiable structure even after being metabolized. That point is illustrated by the aflatoxin B_1(AFB_1) example that we discuss now.

Extrapolation of the carcinogenic effects of aflatoxin B_1(AFB_1) is a good example of comparative process tracing. Produced by certain species of fungi that grow on various types of grains and nuts, aflatoxins are now generally regarded as an important risk factor for liver cancer, a belief dating back to the 1960s that has its origins in laboratory experiments on rats and epidemiological studies (Wogan 1992, 123). Jointly, a positive correlation in epidemiological data between liver cancer and exposure to aflatoxins through food contamination, and a corresponding experimental result in rats, provided a prima facie case for the conclusion that aflatoxins are carcinogenic in humans. However, this evidence alone is not unequivocal. The epidemiological correlation might result in whole or in part from an unmeasured common cause of aflatoxin exposure and liver cancer, while rats might be an inappropriate model for humans with respect to aflatoxins. The appropriateness of the rat as a model in this context was hardly an idle concern, given that aflatoxin was found to have little carcinogenic effect in mice (Gold et al. 1992, 581–82; Hengstler et al. 2003, 491). Differing results among animal models are a clear case of a "reason to suppose otherwise," blocking extrapolation by simple induction. Let us consider how comparative process tracing ameliorated this situation.

Since there are often trans-species differences in the metabolism of foreign compounds, a natural starting point for this inquiry was to analyze the metabolism of aflatoxins in humans and in the rodent populations in which aflatoxins were found to be carcinogenic. It was found that AFB_1, the most common aflatoxin, was converted to the same phase I metabolite across these groups (Wogan 1992, 124). Given the sharp differences in carcinogenic effects of AFB_1 in rats and mice, it was of obvious interest to inquire which of these two animal models was a better guide for humans. It was found that although the phase I metabolism of AFB_1 proceeded similarly among mice, rats, and humans (and in fact at a higher rate in mice), the phase II metabolism among mice was extremely effective in detoxifying AFB_1 but not among rats or humans (Hengstler et al. 1999, 928–31). Furthermore, this metabolite bound to DNA in rat liver cells in vivo at sites at which the nucleotide base guanine was present to form complexes called DNA adducts (ibid., 927). It was further found that such cells suffered unusually frequent mutations in which guanine-cytosine base pairs were replaced with adenine-thymine pairs, a mutagenic effect found in vivo among rats and in vitro among cells of a variety of origins, including bacteria and human (ibid., 923, 927). In addition, guanine-cytosine to adenine-thymine mutations were found in activated oncogenes present in rats exposed to AFB_1 but were absent in the controls (ibid., 130–33). Thus, comparative process tracing yielded the conclusion that the rat was a better model than the mouse.

The example also illustrates that comparative process tracing need not be restricted to comparisons between a single model-target pair, but may involve selecting among several candidate model organisms. In fact, rats and mice were not the only model organisms considered: guinea pigs and hamsters were also studied. These were compared with humans on the basis of quantity of AFB_1 DNA adducts present per unit of peripheral blood among individuals exposed to AFB_1, with a one strain of rat, the Fischer rat, bearing the closest similarity to humans (Hengstler et al. 1999, 925–26). However, even in the Fischer rat, the quantity of DNA adducts was significantly less than in humans, suggesting that even the most sensitive rodent model provides an underestimate of the human impact of AFB_1. The quantity of DNA adducts provides information about a downstream stage of the mechanism (like Z in Figure 5.1). Thus, by focusing on the quantity of DNA adducts, researchers could avoid the cumbersome task of comparing every likely point of difference. This example also demonstrates how comparative process tracing can indicate extrapolative limitations of the best model. In this case, one could reasonably use the Fischer rat to extrapolate the conclusion that AFB_1 exposure increases the chance of liver cancer, and perhaps even use the effect in the Fischer rat to estimate a lower bound for the strength of that effect. But it is doubtful that a quantitative estimate of the impact of AFB_1 upon liver cancer could be correctly extrapolated from the Fischer rat to humans.

5.4 CRITIQUES OF ANIMAL EXTRAPOLATION

An account of extrapolation should be able to adjudicate methodological disputes on this topic, and this section illustrates how the proposal advanced here can do that. In a book and series of articles, Hugh LaFollette and Niall Shanks argue that model organisms cannot be reliably used for extrapolation at all, but only as sources of promising hypotheses to be tested by clinical or epidemiological investigations (1993a, 1993b, 1995, 1996). They use the term *causal analogue model* (CAM) to refer to models that can ground extrapolation, and *hypothetical analogue model* (HAM) to refer to those that function only as sources of new hypotheses. According to LaFollette and Shanks, animal models can be HAMs but not CAMs. A similar though somewhat more moderate thesis is advanced by Weber. He maintains that, except for studies of highly conserved mechanisms, animal models primarily support only "preparative experimentation" and not extrapolation (2005, 185–86). Weber's "preparative experimentation" is similar to LaFollette and Shanks's notion of a HAM, except that it emphasizes the useful research materials and procedures derived from the animal model in addition to hypotheses (2005, 174–76, 182–83). In this section, I argue that these pessimistic claims about the potential of animal extrapolation are not correct.

5.4.1 No Relevant Difference

LaFollette and Shanks's primary argument for the conclusion that model organisms can function only as HAMs and not as CAMs rests on the proposition that if a model is a CAM, then "*there must be no causally relevant disanalogies between the model and the thing being modeled*" (1995, 147; italics in original).[4] It is not difficult to show that animal models rarely if ever meet this stringent requirement. But an obvious reply is that LaFollette and Shanks's criterion of CAM-hood is unreasonably strict. In light of this, LaFollette and Shanks consider the possibility that a weaker condition than the complete absence of relevant causal disanalogies could suffice for extrapolation. They suggest that this proposal be interpreted as follows:

> Begin with two systems, S_1 and S_2. S_1 has causal mechanisms [a, b, c, d, e], S_2 has mechanisms [a, b, c, x, y]. When stimulus s_f is applied to subsystems [a, b, c] of S_1, response r_f regularly occurs. We can therefore infer that were s_f applied to subsystems [a, b, c] of S_2, it is highly probable that r_f would occur. (1995, 153)

However, they argue that this inference is valid only if the relationship between the stimulus and the response is entirely independent of the differing mechanisms, [d, e] and [x, y] (ibid.). But if these mechanisms make no difference to the relationship between the stimulus and the response, then there are no relevant disanalogies between S_1 and S_2, which would mean that S_1 is a CAM after all. Thus, LaFollette and Shanks conclude that when it comes to extrapolation, only a CAM in their sense will do: there must be no relevant causal dissimilarities between model and target (cf. 1996, 180).

Needless to say, this strict condition is rarely if ever satisfied. Not only are relevant differences across species inevitable, but dissimilarities are also extremely common *within species* and even for a *single organism* at different stages of its life. The field of pharmacogenomics, for instance, is dedicated to the study of genetic differences among humans that produce divergent responses to drug therapies. Likewise, susceptibility to, say, harmful side effects of a therapy may be contingent upon factors associated with age, such as declining kidney functioning. Thus, if the strict criterion of CAM-hood proposed by LaFollette and Shanks were accepted, not only would extrapolation from animal to human be illegitimate, but so would extrapolation from humans to other humans. Indeed, even extrapolations from past to future in the life of a single person would be unjustified.[5]

The flaw in LaFollette and Shanks's argument is that it overlooks the connection between the specificity of the claim to be extrapolated and the standard of a suitable model. This point is illustrated nicely by the aflatoxin example. In this case, the Fischer rat would not qualify as a CAM in LaFollette and Shanks's strict sense, since the quantity of DNA adducts resulting from AFB_1 is less in the Fischer rat than in humans. Yet this difference does not undermine extrapolating the positive causal relevance of AFB_1 for liver cancer. The difference suggests that the effect in the Fischer rat is *less* than that in humans. But if the effect in Fischer rats is positive and less than that in humans, then the effect in humans must be positive, too. Consequently, although it would be unwise to extrapolate the *exact* causal effect of AFB_1 upon liver cancer from Fischer rats to humans, the known difference provides no reason against extrapolating a claim about positive causal relevance. Thus, a model might provide a good basis for extrapolating a *qualitative,* but *not a quantitative,* claim concerning a causal effect.

This example suggests that LaFollette and Shanks's stringent criterion of CAM-hood is simply a characterization of what a model organism must be if it is to serve as a basis for the extrapolation of *exact* causal effects. Generally, neither animal-model-to-human nor human-to-human extrapolation can expect such precision. For instance, there is reason to think that the quantitative effect of AFB_1 upon liver cancer varies among human populations. One important reason is that exposure to the hepatitis B virus appears to increase susceptibility to the carcinogenic effects of AFB_1 (cf. Kew 2003), and rates of exposure to that virus vary geographically. LaFollette and Shanks's mistake, therefore, is to present their characterization of a CAM as an entirely general condition required for the extrapolation of any causal claim whatever, when it is in fact only a criterion for extrapolating an extremely precise causal generalization. The conditions that suffice for extrapolating claims concerning positive causal relevance are far less stringent than those needed for extrapolating the exact probability distribution of the effect, conditional on interventions that set the value of the cause.

Chapter 6 explores in greater generality and precision conditions that suffice for extrapolating claims concerning positive or negative causal relevance. In section 6.2.2, I explain how these sufficient conditions are in fact quite reasonable in the aflatoxin example.

5.4.2 The Extrapolator's Circle

LaFollette and Shanks also use the extrapolator's circle as an argument for their conclusion that animal models can function only as HAMs and not as CAMs. They claim, reasonably enough, that the appropriateness of a model organism for extrapolation must be demonstrated by empirical evidence (1993a, 120).[6] But they argue that this appropriateness cannot be established without already knowing what one hopes to learn from the extrapolation.

> We have reason to believe that they [animal model and human] are causally similar only to the extent that we have detailed knowledge of the condition in *both* humans and animals. However, once we have enough information to be confident that the non-human animals are causally similar (and thus, that inferences from one to the other are probable), we likely know most of what the CAM is supposed to reveal. (1995, 157)[7]

LaFollette and Shanks presumably mean to refer to their strict CAM criterion when they write "causally similar," but the extrapolator's circle can be stated independently of that criterion. Whatever the criterion of a good model, the problem is to show that the model satisfies that criterion given only limited, partial information about the target.

However, LaFollette and Shanks's argument shows that extrapolation from animal to human is never legitimate only if it proves the same for extrapolation from one human group to another. For suppose that a particular causal generalization is known to obtain in one human population, and the question is whether it does so in a second. How is one to know whether the two populations are sufficiently similar for the purposes of the extrapolation? According to LaFollette and Shanks, this similarity can be established only on the basis of independently learning the causal relationship in each population and then comparing results. But that would obviate the need for the extrapolation. Thus, the extrapolator's circle shows that animal extrapolation is never justified only if it shows the same about extrapolation in all heterogeneous populations.

This result suggests that the extrapolator's circle does not really show that animal extrapolation can never justify informative conclusions about humans. An account of extrapolation should be able to specify where LaFollette and Shanks's argument goes wrong, while indicating the extent to which the extrapolator's circle is a genuine problem. Unlike previous accounts of extrapolation, the proposal advanced here can do that. LaFollette and Shanks's attempt to turn the extrapolator's circle into a general critique of animal extrapolation overlooks the role of premises concerning likely similarities and differences in analogical reasoning. Thus, in comparative process tracing, providing evidence for the suitability of the model requires comparisons *only* at stages in the mechanism in which significant differences are likely to occur. Consequently, it may be necessary to compare only a few stages of the mechanism. For example, metabolism is the most common source of difference in carcinogenic mechanisms among mammals. Thus, showing that phase I and II metabolism of AFB_1 proceeds similarly in rats and humans strengthens the case for the rat as a model organism. Yet an understanding of the phase I and II metabolism of AFB_1 in humans, considered on its own, provides little information regarding the carcinogenic effects of this compound. Moreover, it is not necessary to compare all points of likely significant difference if there is a downstream stage of the mechanism upon which upstream differences leave their mark. This point is illustrated in the AFB_1 case by the use of the quantity of DNA adducts to assess several potential animal models. In sum, making a case for the suitability of the model may require examining only a few key features of the mechanism in the target, and knowledge of these features alone would fall far short of what one hopes to learn from the extrapolation. In such cases, the extrapolator's circle is avoided.

The extrapolator's circle is a serious challenge if little is known about likely similarities and differences in relevant mechanisms or if it is known that the model and the target are likely to differ in almost every relevant respect. In the latter case, one would effectively know that the organism in question is in fact a very poor model, which would imply that it ought not to be used as a basis for extrapolation. The more interesting case, then, is the first: little is known about likely similarities and differences or their significance for the causal relationship in question. There can be little doubt that such cases sometimes arise, and when they do, extrapolation obviously cannot proceed by comparative process tracing, but would presumably rely upon simple induction. But transforming the extrapolator's circle into a general critique of extrapolation from animal models would require not merely showing that such circumstances *sometimes* arise. It would be necessary to show that this situation is *almost always* the one faced in animal extrapolation. That is an argument that LaFollette and Shanks have not made, and it is one that seems difficult to make, given examples like aflatoxin.

That comparative process tracing can establish the suitability of an animal model also demonstrates that extrapolation is not restricted to entrenched mechanisms inherited from distant ancestors. The carcinogenic mechanism in the AFB_1 example is clearly not of this character since, for instance, it is not present in mice. In short, that a mechanism is highly conserved is *one, but not the only,* possible basis for extrapolation.

5.4.3 HAM Versus CAM?

An underlying assumption of LaFollette and Shanks's argument is that there is a sharp divide between CAMs, which can support extrapolation, and HAMs, which only suggest fruitful hypotheses and lines of research. They write that "there is a big difference between an animal model being a good source of hypotheses and its being a good means to test hypotheses" (1996, 199). LaFollette and Shanks support the claim that there is a strict divide between HAM and CAM by appeal to the old distinction between the contexts of discovery and justification (1996, 194).[8] According to this doctrine, the manner by which a hypothesis is generated has no relevance whatever to the assessment of its scientific adequacy. Whether the new hypothesis was inspired by a dream, a poem, or the floral pattern of a colleague's Hawaiian shirt makes no difference to its epistemic virtues, which can be decided only through a careful examination of the relevant evidence. The sharp contrast between HAM and CAM drawn by LaFollette and Shanks is simply the context of discovery versus justification distinction applied to animal models. HAMs are animal models in the context of discovery, while CAMs are models in the context of justification.

However, the context of discovery versus justification dichotomy has been critiqued from a wide variety of perspectives (cf. Hanson 1958; Kuhn 1977, chap. 11; Longino 1990; Darden 1991; Kelly 1996; Simon 1998).

Current discussions of the distinction in the philosophy of science literature take it as more or less given that aspects of the discovery process can be relevant to the assessment of hypotheses, and then proceed to consider the finer points of proposals about how this is so (cf. Darden and Craver 2002; Castle 2001; Elliott 2004). The problem with the thesis that there is an unbridgeable chasm between the contexts of discovery and justification can be appreciated by means of simple examples like the following. Imagine two procedures for generating hypotheses, the first of which generates correct hypotheses 95 percent of the time and the second that generates correct hypotheses 1 percent of the time. Now suppose that the two procedures have produced conflicting hypotheses. Given this information, which hypothesis—the one generated by the first procedure or the one generated by the second—do you think is more likely to be correct?

The obvious answer is the hypothesis produced by the first procedure. Thus, that a hypothesis was generated by a procedure likely to produce empirically successful hypotheses can be relevant evidence. Although it is rarely possible to assign exact rates of success to distinct discovery procedures, the process is nevertheless typically not a matter of ineffable and mysterious inspiration either. For example, scientific discovery is typically guided by prior knowledge of constraints that must be satisfied by a successful hypothesis in the domain in question. Ignoring these constraints is likely to lead to a grossly inadequate hypothesis. In sum, the process by which hypotheses are discovered is amenable to logical analysis and can be relevant evidence to be considered in assessing the hypothesis.

A defender of the context of discovery versus justification dichotomy might object that the mode of discovery is evidentially relevant only insofar as it suggests that the hypothesis is consistent with particular observations or experimental results. Consequently, the mode of discovery would be irrelevant to one who knew all of these data and who was able to directly assess the hypothesis with regard to them. That may be true, but it is nevertheless the case that information about the mode of discovery may be evidentially relevant to someone in a less than perfect epistemic position. One might not know what all of the relevant data are, or one might not be able to directly assess whether the hypothesis is consistent with them. In such cases, information regarding the source of the hypothesis may remain evidentially relevant. This is very much the situation one faces with regard to animal extrapolation. For instance, to a person with complete knowledge of carcinogenesis in humans, information about animal models would be irrelevant for assessing the accuracy of any hypothesis about the effects of AFB_1. But for ordinary mortals who lack such perfect knowledge, animal models can be a useful source of evidence.

These considerations are directly relevant to the supposed sharp distinction between HAM and CAM. As LaFollette and Shanks observe, although hypotheses can be inspired by practically anything, not everything is a good HAM (1996, 195). The most obvious way a model could be a good HAM is in virtue of being likely to generate hypotheses about the target that are true, or at least approximately so. Yet this account of what makes a good HAM entails that the difference between HAM and CAM is one of degree. Both provide some evidence for the extrapolation; it is just that the evidence provided by the CAM is stronger and less equivocal. But LaFollette and Shanks cannot distinguish between good and bad HAMs in this manner, since that would contradict their claim that *only* CAMs in their very strict sense provide *any* evidence for extrapolation.

So what does make a good HAM, according to LaFollette and Shanks? They write, "A HAM is likely to be valuable if there are demonstrable functional similarities between the model and item modeled" (1996, 195). But it is difficult to see how this could be true, given their persistent claim that functional similarity is no indicator of similarity of mechanisms.[9] For in that case, there is no reason to think that a functionally similar HAM will lead to fruitful hypotheses rather than unproductive dead ends. Of course, model organisms typically share more with their targets than mere functional similarity. They also share a common ancestor and some fundamental mechanisms at the level of biochemistry, the cell, and physiology. These similarities provide some—albeit rather uncertain and rough—grounds for extrapolation. And that is what justifies regarding them as HAMs.

Rather than a sharp dichotomy between HAMs and CAMs, then, there is a continuum from models providing weaker to those providing stronger grounds for extrapolation. A model might be a weak basis for extrapolation because little is known about likely sources of significant difference and similarity, or because mechanisms in the model and the target have not been compared at stages of likely difference. The more that is known about likely similarities and differences, and the more the likely differences have been checked and found to be absent, the stronger the basis for extrapolation. Moreover, exactly how similar the model is required to be depends upon the claim of interest to the extrapolation, as noted above and explored in further detail in Chapter 6. From this perspective, any sharp HAM versus CAM distinction is inevitably arbitrary and ultimately unimportant. The pertinent issues are how thoroughly comparative process tracing has been carried out and what conditions are required to extrapolate the generalization in question.

Despite disagreeing with LaFollette and Shanks's methodological critique of animal extrapolation, I think that they deserve credit for articulating objections that had not been adequately addressed in the literature on this topic. I also think that they are correct that these methodological questions matter to ethical issues surrounding animal experimentation, since animal research is typically justified on the grounds that it provides knowledge that benefits humans. Thus, the ethical question turns on whether the benefit to humans outweighs the suffering of the animal model. Although an in-depth exploration of these ethical issues is beyond the scope of this book, I would like to briefly indicate what I regard as the main ethical implication of the account of extrapolation developed here. LaFollette and Shanks wish to argue that animal experimentation is unethical in general, and hence they endeavor to show that extrapolation, in general, is not a reliable source of new information about humans. And if animal extrapolation were indeed so utterly incapable of providing useful information concerning humans, then the standard ethical defense of animal research would be undermined. In contrast, I suggest that extrapolation is reliable and informative in some circumstances but not others, and make some steps toward clarifying what those circumstances are. This perspective calls across-the-board moral vindications or condemnations of animal research into question.[10] Whether animal research is ethically defensible in a given case may depend *in part* upon the potential for extrapolating useful information about humans. And the extent to which this is or is not possible will depend on complex, case-specific scientific details. I do not pretend to answer the question of whether and to what extent animal research is ethically defensible. However, I do think that my account of extrapolation casts doubt on any "one size fits all" argument on either side of the issue.

5.5 CONCLUSION

This chapter presents a mechanisms approach that addresses some of the primary methodological challenges confronting animal extrapolation. I began by considering simple induction, which is an undeniably important aspect of extrapolation but also limited in important ways. Simple induction alone would result in many mistaken extrapolations from animals to humans. In addition, there often is some reason to suppose that the extrapolation might be inaccurate, and simple induction provides little guidance about what to do when that is the case. More sophisticated approaches to extrapolation attempt to indicate how the suitability of a model for a particular extrapolation could be established. Any proposal of this sort must confront what I called the extrapolator's circle. That is, it must explain how the suitability of the model could be established without already knowing what the extrapolation is supposed to tell us. Moreover, since causally relevant disanalogies between animal models and human targets are inevitable, it is necessary to explain how extrapolation can be legitimate even when such disanalogies are present. I argued that existing proposals concerning extrapolation—either in terms of capacities or in terms of mechanisms—fail to adequately address either of these challenges. However, I proposed that the mechanisms approach can be developed so as to provide an answer to the extrapolator's circle. The key proposition in this proposal is what I called comparative process tracing. Comparative process tracing depends upon possessing information about the stages at which significant differences in mechanisms are and are not likely to occur, and on the directional property of the mechanism which enables one to focus on downstream stages when looking for relevant difference. Thus, it may be possible to establish the suitability of a model organism through a comparison with the target at a small number of stages in the mechanism. Finally, I examined several general methodological objections to animal extrapolation that were motivated by concerns about the ethical permissibility of animal research from the perspective of the approach to extrapolation proposed in this chapter. Although I think that these objections raise important issues, I argued that they are unsuccessful. In the next chapter, I explore conditions that can justify extrapolating claims of positive or negative causal relevance in greater detail, and suggest that this topic is closely relevant to the issue of ceteris paribus laws.

6

Ceteris Paribus and Extrapolation

Laws and generalizations qualified by the expression "*ceteris paribus*," a Latin phrase for "other things being equal," are argued by some to play an important role in biology and social science. In contrast, others object that there is no satisfactory interpretation of ceteris paribus (hereafter, cp) laws and that they are not useful for understanding characteristic generalizations in the biological or social sciences. This chapter examines the controversy over cp laws from the perspective of extrapolating claims about positive or negative causal relevance. I propose that considering the topic in this light helps to resolve a central puzzle concerning the scientific role of cp laws.

A survey of the current literature on the topic reveals that the expression "cp law" is highly ambiguous: several types of generalizations have been classified under this label. This point has been made explicitly by Gerhard Schurz (2001b, 2002), and it is also implicit in the variety of proposals concerning the manner in which cp laws should be understood. On some of these interpretations, cp laws are in fact illustrated by causal claims encountered in earlier chapters, such as causal effects and descriptions of mechanisms. The issue of cp laws is also related to extrapolation: one might say that a causal generalization found in one context will also obtain in another, provided that *nothing interferes* or *all else being equal*. That is, the expression "ceteris paribus" can serve as a vague, all-purpose term for indicating whatever conditions are needed for the extrapolation to be correct. Moreover, extrapolation is an important part of what motivates discussions of cp laws, since the content of the cp clause is intended to provide guidance about when the generalization can and cannot be appropriately applied.

A common type of analysis of cp laws known as the "completer approach" interprets laws as universal generalizations and the cp clause as stating conditions in which the law holds without exception. But in cases in which the conditions that lead to exceptions to the law cannot be listed exhaustively, the completer approach inevitably violates what I call the *domain specificity criterion*. This criterion requires that a law of a domain should provide information specifically about that domain rather than merely asserting, say, that determinism is true. I propose that the failings of the completer approach arise from two sources. First, it interprets "ceteris paribus" as qualifying a *generalization* in cases in which that expression should be understood in reference to an *inference schema*. Unlike an empirical law, an inference schema (such as *modus tollens*) need not provide domain-specific information. Second, the completer approach presumes that the generalization in question is a universally quantified sentence, typically of the form "All Fs are Gs." When "cp" is taken to indicate an inference schema that specifies conditions for extrapolating a claim about causal relevance, the problems afflicting the completer approach disappear.

Making this case requires a more detailed account of the conditions that suffice for the extrapolation of claims of positive or negative causal relevance. Relying on groundwork of earlier chapters, such an account is provided in section 6.2. The sufficient conditions in question are articulated in what I call the *extrapolation theorem*. I discuss some ways in which the scope of the extrapolation theorem can be extended, and illustrate its application by means of the aflatoxin example introduced in Chapter 5. The extrapolation theorem reinforces the claim made in Chapter 5 that similarity in *all* causally relevant respects is *not* necessary for a model to serve as a basis for extrapolating claims about positive or negative causal relevance.

6.1 THE MANY MEANINGS OF CETERIS PARIBUS

The ambiguity of the expression "cp law" is important to the present discussion, since on some interpretations "cp law" refers to a relatively unproblematic type of generalization while the opposite is true for other interpretations. I begin with the less problematic kinds, and then turn to the particularly troublesome ones, which I group under the heading "completer approach."

6.1.1 Comparative, Normative, and Definite

A striking feature of the philosophical literature on cp laws is the variety of types of generalization that are referred to by that label. The best classification of interpretations of "cp law" that I know of is due to Schurz (2001b, 2002). Schurz divides cp laws into two main types, *exclusive* and *comparative*. An exclusive cp clause indicates an absence of factors that would produce exceptions to the law, whereas a comparative cp clause asserts not that interfering factors are absent but that they are distributed identically between groups that differ with respect to the putative cause. Thus, the comparative sense of cp can be satisfied while the exclusive sense is not, for example, if there is a disturbing factor that is distributed identically in both groups. Causal effects and qualitative descriptions of them, such as claims about positive causal relevance, are examples of generalizations that fall into Schurz's category of comparative cp laws. Since there is a well-established procedure for estimating causal effects, namely, the randomized controlled experiment, there is little plausibility in the claim that such generalizations are somehow scientifically illegitimate. In such disciplines as macroeconomics or evolutionary biology, wherein controlled experiments are often not a practical possibility, there is a genuine *epistemological* challenge of estimating causal effects. But the fact that in some circumstances it is difficult to ascertain the truth of a particular type of generalization is no reason for claiming that such generalizations are meaningless or unworthy of science.

Although it is less frequently encountered in the philosophical literature than the exclusive interpretation, the comparative interpretation of cp laws does have some proponents (cf. Morreau 1999). In contrast, Woodward regards claims about positive causal relevance as a kind of generalization that cannot be adequately interpreted as cp laws, which he presumes must be understood in the exclusive sense (2002b, 306–16).[1] Whether causal effects should count as cp laws or whether only generalizations falling in the exclusive category truly deserve that title is, in my judgment, an uninteresting terminological quibble. Nevertheless, there are two important points to be made about the distinction between comparative and exclusive cp laws. First, the comparative interpretation illustrates that some types of generalizations that philosophers and others have in mind when they use the term "cp law" are relatively unproblematic. This helps explain the incredulous reaction of many to arguments that cp laws are meaningless, untestable, and so on, and entails that there is a sense in which such critiques are certainly mistaken.[2] The second important point is that an account of how *comparative* cp laws can be tested and used in science cannot be called upon in defense of *exclusive* cp laws. For example, it is not rare to find the scientific legitimacy of exclusive cp laws defended on the grounds that they can be tested by randomized controlled experiments (cf. Hausman 1992, 137; Kincaid 1996, 67–68). But, as will be explained in the next subsection, controlled experiments cannot be used to test the most problematic sort of exclusive cp laws.

Schurz subdivides exclusive cp laws into three types: *normic*, *definite*, and *indefinite*. Normic exclusive cp laws are exemplified by simple generalizations such as "Birds have wings." This is normally true, although some birds may be wingless owing to mutation or amputation. Descriptions of mechanisms are another, more scientifically interesting example of normic generalizations: the description of the HIV replication mechanism given in Chapter 4 is an account of how this process *normally* transpires. Marc Lange (1993, 2000, 2002) proposes an account of cp laws that I interpret as falling into the normic exclusive category. At the heart of Lange's general account of laws of nature is what he terms the *root commitment*: laws are the most reliable rules for making inferences in a specified domain (2000, 23–28). In the case of a cp law, the qualifying clause "need not refer to the *complete* list of influences in order for the law to be (in the relevant range of cases) accurate enough for its intended purposes" (2000, 175). Rather, the cp clause includes "all of the other influences great enough in such cases to be nonnegligible for certain purposes" (ibid., 175). So, when the conditions specified in the cp clause hold true, the law is normally accurate enough for the purposes that it is intended to serve, though it may break down in some unusual circumstances.

Similarly, Schurz argues that it is natural to interpret normic exclusive cp laws in such a way that accepting one implies an endorsement of a default inference (cf. Schurz 2001a, 2001b).[3] For example, when you learn that something is a bird, it is justifiable to make the default assumption that it also has wings, a conclusion that may be revoked upon acquiring additional information. Evidently, this default inference is reasonable only if it is the case that most birds have wings. Hence, on Schurz's account the normic interpretation of cp laws is closely tied to a statistical condition: if it is a normic exclusive law that under conditions C, As are Bs, then most As are Bs when conditions C obtain. The link between normic exclusive cp laws and statistical regularities makes it easy to understand how such generalizations can be supported or undermined by data. Thus, like comparative cp laws, normic exclusive cp laws are relatively unproblematic. Having implications about what is usually the case also distinguishes normic exclusive cp laws from the definite and indefinite exclusive varieties. A law that holds under ideal conditions or when nothing interferes may obtain only rarely if the ideal conditions are usually not approximated or if interfering factors are ubiquitous.

Definite exclusive cp laws are illustrated by such examples as the law of the pendulum or Galileo's law of free fall: one can specify ideal conditions in which the laws are true without exception. Presumably with such examples in mind, Cartwright characterizes cp laws as "laws that hold under special conditions, usually ideal conditions" (1983, 45). Like the comparative and normic exclusive varieties, definite exclusive cp laws are not particularly troublesome. The law can be tested if one can approximate the ideal conditions in a laboratory setting, for instance.

A single generalization might be interpreted as a comparative or as a normic or definite exclusive cp law, depending on the context. For instance, the claim that increasing the supply of a commodity leads to a reduction in its price could be understood as a claim about negative causal relevance in a particular population (a comparative cp law). Or it could be understood as asserting that this relationship *normally* obtains across some collection of populations. Or it could be understood in reference to theorems that specify ideal conditions in which the laws of supply and demand hold without exception. And there are contexts in which each sort of generalization would be useful. Knowing the relationship between, say, the supply and price of oil in a particular economy can obviously be of great practical importance. The knowledge that increases in supply normally produce decreases in price is valuable when considering the effects of changes in supply in a new situation. And a precise specification of ideal conditions in which the laws of supply and demand obtain can be useful for explaining exceptions to the usual pattern. The fact that generalizations with a cp clause attached can mean so many different things helps one understand why there would be confusion and dissent about what such claims assert.

6.1.2 The Completer Approach

Indefinite exclusive cp laws are far more problematic than any of the three types of cp laws considered above. Cp laws of this sort are *exclusive* in virtue of asserting that the law holds so long as nothing interferes, and they are *indefinite* insofar as one is not able to specify a set of conditions in which nothing interferes and hence in which the law definitely holds. Furthermore, they differ from normic exclusives, since the cp clause is intended to specify conditions in which the law holds not just for the most part but *without exception*. Several analyses of the truth conditions of indefinite exclusive cp laws exist (cf. Fodor 1991; Hausman 1992; Pietroski and Rey 1995). They can be collectively dubbed the "completer approach,"[4] as each proposal attempts to identify an appropriate way to characterize conditions, typically labeled *C*, that complete the law, that is, in which the law is true without exception. Since the cp law is presumed to be indefinite, it is not possible to exhaustively list the factors capable of interfering with the relationship in question; thus, *C* must be specified in some less direct manner. The chief difficulty with existing versions of the completer approach is that they inevitably violate what I call the *domain specificity requirement*, according to which laws of a domain should provide information specifically about it. For example, laws of economics should provide information about economic phenomena and not merely assert a logical truth or a proposition about metaphysics.

The simplest version of the completer approach is to construe the completing clause as a negated existential that quantifies over all possible things that could prevent the occurrence of the outcome despite the presence of the cause. For instance, if the cp law is of the form "Cp, all Fs are Gs," then the completed form on this proposal is "Anything that is F will also be a G unless some factor is present that causes it not to be a G." Yet as Schurz (2001b) shows, this sort of proposal makes cp laws *almost empty*, in the sense that "Cp, all Fs are Gs" is equivalent to the claim that for everything that is an F, there are deterministic causes of whether it is a G. Moreover, if there are deterministic causes of whether something is G, then there are deterministic causes of whether it is *not* G. Hence, the negated existential version of the completer approach has the highly undesirable consequence that "Cp, all Fs are Gs" entails "Cp, all Fs are not Gs."[5]

The negated existential interpretation illustrates in a particularly striking way the fundamental problem that confronts all existing variants of the completer approach: they all make "Cp, all Fs are Gs" equivalent to claims asserting the existence of deterministic causes of G conditional on a thing's being F (Schurz 2002, 354–64). Distinct versions of the completer approach differ only with regard to slight variations in what features those deterministic causes are required to satisfy. For instance, causes that prevent an F from being G must be independently identifiable (Pietroski and Rey 1995), or for every realization of F, there must be a cause C such that F and C are jointly but not separately sufficient for F (Fodor 1991).[6] What is problematic about these definitions is that they interpret cp laws as claims that provide no information specifically relevant to the intended subject matter of the supposed law. That is, the completer approach violates the following criterion.

> *Domain Specificity Criterion*: Laws should provide information specifically relevant to their intended domain of application.

For example, the generalization that raising interest rates slows inflation is intended to provide information about the relationship between these quantities in economic contexts, and not merely to assert that there are deterministic causes of inflation. For there may be deterministic causes of inflation even if interest rates cannot be used to predict inflation, cannot be used to control inflation, and so on.

The domain specificity criterion seems quite obvious and unassailable. Indeed, it is similar to Lange's root commitment, according to which the laws of a domain are the best rules of inference in that domain. Clearly, a generalization that provides no specific information about a domain cannot qualify as one of its laws according to Lange's criterion. Moreover, the domain specificity criterion is a reasonable requirement for any generalization that is intended to express important knowledge characteristic of some domain, whether or not it is graced with the honorific title "law." Important generalizations in HIV research, for instance, aim to provide information about how the virus replicates, its effects on various features of the immune system, and so forth. Whether one judges such generalizations to be *laws*, it is clear that they ought to satisfy the domain specificity criterion.

In addition to running afoul of the domain specificity criterion, the completer approach defines cp laws in such a way as to make it very hard to understand how empirical evidence could provide reason to accept or reject them. That is, the completer approach transforms cp laws into claims about the existence of deterministic causes of various sorts. Yet determinism is a metaphysical doctrine that one can hardly hope to settle by investigations in such fields as economics or molecular biology. Typically, advocates of the completer approach say very little about how cp laws are to be tested. But one exception is Daniel Hausman, who defines cp laws as follows:

> A sentence with the form, "*Ceteris paribus* everything that is an *F* is a *G*" is a law just in case the *ceteris paribus* clause determines a property *C* in the given context, and it is a law that everything that is *C* and *F* is also *G*. (1992, 136)

Hausman does not explain how the cp clause "determines" a completer C in a given context, nor does he say much about what such a completing clause would look like. Consequently, it is unclear how Hausman's proposal differs from interpreting "Cp, all Fs are Gs" to mean that for anything that is F, there are further causes that would suffice to make it G. Thus, cp laws as defined by Hausman appear to violate the domain specificity condition, as do the other versions of the completer approach. But let us consider how cp laws can, according to Hausman, be supported or undermined by empirical evidence.

Hausman states that cp laws can be tested by controlled experiments (ibid., 139). Yet it is clear that this is not true if cp laws are interpreted as he proposes, or according to any other version of the completer approach. For on every version of the completer approach, cp laws are equivalent to claims concerning the existence of deterministic causes, but it is obvious that determinism is not a matter that can be settled by a randomized controlled experiment. For example, consider a clinical trial in which the rate of recovery is significantly higher among those who received the treatment than in the control group. If the experiment was properly designed and implemented, this result would support the claim that the treatment is a positive causal factor for recovery, at least within the population represented by the sample. However, the experimental result in no way demonstrates that there is a law of nature of the form "Whenever the treatment is present in conjunction with the condition C, recovery invariably ensues." The experimental result is consistent with the world being fundamentally indeterministic, and hence with the complete absence of deterministic laws of nature. I suspect that the claim that cp laws, as construed by the completer approach, can be tested by randomized controlled experiments results from failing to distinguish the comparative and exclusive senses of "cp."

Hausman also proposes four standards that a cp law must meet if it is to be judged acceptable; a cp law candidate must be "lawlike, reliable, refinable, and excusable" (ibid., 141). Lawlike generalizations can support counterfactuals, are confirmed by their instances, and can be used in explanations (ibid., 29–93). A generalization is said by Hausman to be reliable just in case there is some class of cases in which it usually holds even if the cp clause is ignored (ibid., 141). The generalization is refinable when this class of cases in which the generalization is reliable, sans cp clause, can be extended through the addition of "qualifications" (ibid., 140–41). Finally, the generalization is excusable if specific factors can be identified to account for its failures.

The features described by Hausman are certainly desirable ones for a generalization to have. However, there is little connection between a generalization satisfying the above requirements and its being a cp law in Hausman's sense, or in the sense of the other versions of the completer approach. Suppose that there is a condition C such that it is a law of nature that C and F are sufficient for G. It does not thereby follow that that "All Fs are Gs" supports counterfactuals, for even if you had been F, you might have been in a condition other than C. Nor does it follow that the generalization "All Fs are Gs" is reliable in Hausman's sense, since most

Fs may fail to be Gs if the condition C rarely obtains. In contrast, an exclusive normic cp law could support counterfactuals and be reliable and refinable, and it might do so in a fundamentally indeterministic world in which there are no universal laws of the sort required by Hausman's proposal.

Determinism seems relevant only to the last of Hausman's four criteria. If the world is deterministic, then there is always some explanation of why the generalization failed to obtain in a given case. In an indeterministic world, there might arise two cases identical in all relevant respects, except that the generalization was correct in one and not in the other. However, with respect to complex systems such as an economy or an organism, it is extremely rare that a pair of cases is identical in all relevant respects. Moreover, even if determinism is true, one typically does not know all of the relevant factors that may be responsible for accounting for why a generalization held in one case rather than another. Defining cp laws in the fashion of the completer approach, then, commits one to hauling around some heavy but not very useful metaphysical baggage.

In the remainder of this chapter, I suggest a distinct interpretation of indefinite exclusive cp clauses that focuses on extrapolating positive or negative causal relevance. That proposal requires some further elaboration about the circumstances that license the extrapolation of probabilistic causal claims.

6.2. EXTRAPOLATING PROBABILISTIC CAUSAL CLAIMS

Chapter 5 described how comparative process tracing can be used to extrapolate a mechanism from a model organism to a target. Establishing the existence of a mechanism from cause to effect in the target population is a significant step in extrapolation, yet this alone often fails to tell us much of what we would like to know. Given the disruption principle, the presence of a mechanism licenses the conclusion that interventions on the cause make a difference to the probability of the effect. However, this does not tell us *how* this probability is changed—for instance, whether the probability of the effect is increased or decreased—nor does it provide information concerning the strength of that effect.[7] In the next two subsections, I utilize the conceptual apparatus presented in earlier chapters to develop a mechanisms approach to the extrapolation of claims of positive or negative causal relevance. In section 6.2.1, I use the disruption principle to derive some useful equations linking mechanisms and causal effects (see equations (6.8) and (6.9)). These equations serve as the basis of the proposals advanced in the following subsection. In section 6.2.2, I introduce a concept that I dub *consonance*, according to which different combinations of mechanisms do not exert conflicting positive and negative influences. Given consonance, I prove what I call the extrapolation theorem, which specifies a set of sufficient conditions for extrapolating a causal claim about positive or negative relevance. I then discuss some ways in which the scope of the extrapolation theorem can be broadened and explain how the conditions of the theorem are quite plausible in the example of aflatoxin B_1 and liver cancer, which was discussed in Chapter 5.

6.2.1. From Mechanisms to Causal Effects

The disruption principle, introduced in Chapter 4, provides a connection between mechanisms and causal effects: interventions on X make a difference to the probability distribution of Y just in case there is an undisrupted mechanism from X to Y in the population. But as useful as the disruption principle is, it is clear that a more detailed connection between mechanisms and causal effects is required if the mechanisms approach to extrapolation is to bear much fruit. In this section, I further develop the framework expounded in earlier chapters for this purpose.

As before, let $\mathbf{M}_{XY}$ be the set of mechanisms from X to Y in the population of concern. Let $\wp(\mathbf{M}_{XY})$ be the power set of $\mathbf{M}_{XY}$ (i.e., the set of all subsets of $\mathbf{M}_{XY}$). For example, if $\mathbf{M}_{XY}$ is $\{M_1,M_2\}$, then $\wp(\mathbf{M}_{XY})$ is $\{\emptyset,\{M_1\},\{M_2\},\{M_1,M_2\}\}$. Since $\mathbf{M}_{XY}$ is finite, $\wp(\mathbf{M}_{XY})$ is, too, which means that the members of $\wp(\mathbf{M}_{XY})$ can be numbered 0, 1, ..., n. Although which members of $\wp(\mathbf{M}_{XY})$ are assigned which numbers is immaterial, I assume for convenience that 0 always designates the empty set. For example, the members of $\{\emptyset,\{M_1\},\{M_2\},\{M_1,M_2\}\}$ might be numbered 0, 1, 2, and 3, respectively. Let Φ_i denote the subset of individuals in the population who contain exactly those mechanisms in the ith member of $\wp(\mathbf{M}_{XY})$. In the present example, Φ_0 would be the subset of exactly those individuals who possess no undisrupted mechanisms from X to Y; Φ_1, the subset of those who possess only mechanism M_1 in an undisrupted state; and so on.

The formalism laid out in the above paragraph enables us to discuss partitions of the population according to the presence and absence of the different combinations of mechanisms. The usefulness of this can be seen from the following familiar theorem of probability (cf. Stirzaker 2003, 128–29):

$$E(Y) = \sum_{i=0}^{n} P(\Phi_i)E(Y \mid \Phi_i) \tag{6.1}$$

where 0, 1, ..., n is a partition (that is, mutually exclusive and collectively exhaustive).

In the present case, the Φ_is indicate which combination of undisrupted mechanisms the individual possesses, as explained in the preceding paragraph. For ease of notation, let $E_i(Y) =_{df} E(Y \mid \Phi_i)$, which is to say the expected value of Y in Φ_i, the subset of the population consisting exactly of those individuals who possess all and only the mechanisms in the ith member of $\wp(\mathbf{M}_{XY})$. Given this notational convenience, (6.1) can be abbreviated:

$$E(Y) = \sum_{i=0}^{n} P(\Phi_i)E_i(Y) \tag{6.2}$$

Yet our interest lies not so much with the expected value, $E(Y)$, but with the average causal effect, $E(Y \mid do(x))$, which tells us how the expected value of Y changes with interventions on X. But from (6.2), an equation for the average causal effect can be easily derived simply by conditioning on $do(x)$.

$$E(Y \mid do(x)) = \sum_{i=0}^{n} P(\Phi_i \mid do(x))E_i(Y \mid do(x)) \tag{6.3}$$

That is, (6.3) specifies the average causal effect in the entire population in terms of the sums of the products of the average effect in each cell and the probability of that cell, given the intervention.

Equation (6.3) can be derived for any partition of the population whatever—the fact that the partition is in terms of which combination of mechanisms is possessed by the individual has played no role so far. But it plays an essential role in a crucial simplification of (6.3). Consider $P(\Phi_i \mid do(x))$, the probability that the individual possesses all and only the mechanisms in the ith member of $\wp(\mathbf{M}_{XY})$ conditional on an ideal intervention that fixes the value of X. Recall from definition 2.1 that an ideal intervention eliminates other influences upon X but otherwise makes no changes to the causal relationships. In particular, an ideal intervention on X will not eliminate or add any causal paths emanating from X. Consequently, since Φ_i indicates the combination of mechanisms from X to Y present in the individual, $P(\Phi_i \mid do(x)) = P(\Phi_i)$. Notice that this probabilistic independence need not obtain for all possible partitions of the population; in particular, it will not hold when one partitions by properties that are effects of X. The premise $P(\Phi_i \mid do(x)) = P(\Phi_i)$ allows equation (6.3) to be simplified to this especially useful form.

$$E(Y \mid do(x)) = \sum_{i=0}^{n} P(\Phi_i)E_i(Y \mid do(x)) \tag{6.4}$$

Let $\Delta E(Y \mid do(x))$ be defined as $E(Y \mid do(x)) - E(Y \mid do(x_0))$, where, as before, x_0 is a comparative value of X that is smaller than every value of X in the interval ϑ of concern (see section 2.3.2). From equation (6.4), $E_i(Y \mid do(x))$ and $E_i(Y \mid do(x_0))$ equal

$$E(Y \mid do(x)) = \sum_{i=0}^{n} P(\Phi_i)E_i(Y \mid do(x))$$

$$E(Y \mid do(x_0)) = \sum_{i=0}^{n} P(\Phi_i)E_i(Y \mid do(x_0)) \tag{6.5}$$

Subtracting the bottom equation from the top one and collecting terms gives

$$\Delta E(Y \mid do(x)) = \sum_{i=0}^{n} P(\Phi_i)\Delta E_i(Y \mid do(x)) \tag{6.6}$$

It is natural to identify the probability of a given combination of mechanisms with its relative frequency in the population, since the most straightforward way to understand this probability is as the chance that an individual chosen at random from the population would possess just that combination of mechanisms. Let φ_i be the relative frequency of Φ_i. Hence, in the example described above, φ_0 would be the relative frequency of those who possess no undisrupted mechanism from X to Y; φ_1, the proportion of those who possess only M_1 in an undisrupted state; and so forth.

Thus, equation (6.6) can be rewritten as

$$\Delta E(Y \mid do(x)) = \varphi_0 \Delta E_0(Y \mid do(x)) + \sum_{i=1}^{n} \varphi_i \Delta E_i(Y \mid do(x)) \tag{6.7}$$

Recall that 0 indicates that no undisrupted mechanisms are present in the individual. Consequently, the disruption principle entails that the first term on the right-hand side of (6.7) is zero—that is, since there are no undisrupted mechanisms from X to Y in Φ_0, interventions on X make no difference to the probability distribution of Y within that subpopulation. Equation (6.7) therefore simplifies to the following:

$$\Delta E(Y \mid do(x)) = \sum_{i=1}^{n} \varphi_i \Delta E_i(Y \mid do(x)) \tag{6.8}$$

In the special case in which X and Y are both binary variables, $\Delta E(Y \mid do(x))$ equals $P(Y = 1 \mid do(X = 1)) - P(Y = 1 \mid do(X = 0)) =_{df} \Delta P$. Hence, in this case equation (6.8) takes the following concise form:

$$\Delta P = \sum_{i=1}^{n} \varphi_i \Delta P_i \tag{6.9}$$

Equations (6.8) and (6.9), which were derived from (6.4), the disruption principle, and the definition of an ideal intervention, are the basis of the extrapolation theorem demonstrated in the following section.

6.2.2 Consonance and Causal Relevance

Consider again the example in which it is desired to know whether a particular substance found by experiment to be carcinogenic in rats is also a carcinogen in humans. This case can be phrased in the abstract as follows: X is known to be a positive causal factor for Y in one population, and we want to know whether it is also such in another. In this section I introduce a circumstance, which I dub *consonance*, that greatly facilitates judgments concerning extrapolation. The best way to grasp this concept is by means of an example in which it is *not* satisfied. Consider the classic example of birth control pills and thrombosis: the pills promote the illness but also prevent pregnancy, which is itself a cause of thrombosis (cf. Hesslow 1976; Cartwright 1989, 99–103). Since pregnancy is a more effective cause of thrombosis than oral contraceptives, the pills have a net preventive effect when both mechanisms are at work. However, among women who stand no chance of becoming pregnant, the pills would be positively relevant to thrombosis.

In contrast, consonance holds when distinct combinations of mechanisms do *not* exert conflicting positive and negative influences. More precisely, the mechanism set $\mathbf{M}_{XY}$ is *positively consonant* with respect to the population P just in case there is no subpopulation Φ_i of P such that $\varphi_i > 0$ and X is a negative causal factor for Y in Φ_i. The term *negatively consonant* is defined in the same manner, but switching "negative" and "positive." A mechanism set is *consonant* just in case it is positively or negatively consonant. Recall that Φ_i is the subpopulation of individuals in P who possess exactly those mechanisms in the ith member of $\wp(\mathbf{M}_{XY})$, the power set of $\mathbf{M}_{XY}$, and that φ_i is the relative frequency of Φ_i. Thus, positive consonance asserts that X is not negatively relevant to Y for any combination of mechanisms found in the population.

Consonance is similar but not equivalent to a condition sometimes called "balance" (cf. Selten 2001, 31–32). A mechanism set is *positively balanced* if each mechanism individually exerts a positive influence upon the effect; *negatively balanced* if every mechanism exerts a negative effect; and *unbalanced* otherwise. There are circumstances in which consonance entails balance, namely, when (a) $\varphi_i > 0$ for each Φ_i ; (b) every causally relevant factor is either positively or negatively relevant; and (c) the disruption principle is true. To see that positive consonance entails positive balance when (a)–(c) hold, consider an arbitrary mechanism in $\mathbf{M}_{XY}$. Let Φ_a be the subpopulation of P consisting of individuals who possess only this mechanism from X to Y, and let φ_a be the relative frequency of Φ_a in P. By (a), φ_a is greater than zero. Thus, given positive consonance, X is not a negative causal factor for Y in Φ_a. But then from (b) it follows that X is either a positive causal factor for Y or not causally relevant. Yet since there is a mechanism from X to Y in Φ_a, the disruption principle rules out the second of these two possibilities.

However, conditions (a) through (c) do not suffice for balance to entail consonance. The reason for this is that positive balance requires only that each mechanism acting individually exert a positive influence, but is silent about what occurs when two or more mechanisms operate in tandem. It is possible that two mechanisms that promote an effect separately have the opposite effect when operating jointly. For example, two chemicals might each separately tend to relieve headaches but interact so as to cause headaches when taken together. Balance would entail consonance if an additional assumption were added to (a) through (c) to rule out such contrary interactions, for instance, that the joint effect of two or more mechanisms is always the sum of effects in isolation.

Furthermore, consonance may be true, yet balance false if condition (a) does not obtain. Imagine that there are conflicting causal paths, a stronger positive one and a weaker negative one. In this case, owing to the existence of counteracting causal pathways, the mechanism set is unbalanced. But if the negative mechanism occurs only in conjunction with the positive mechanism, positive consonance can still obtain. That is, in a subpopulation in which only the negative mechanism was present, the cause would tend to prevent the effect. Yet consonance could nevertheless be true if the relative frequency of this negative-mechanism-only subpopulation is zero.

Let us consider how consonance can facilitate extrapolation. Suppose that X is a binary variable that indicates whether or not the individual was exposed to a particular substance, and Y is a binary variable indicating whether the individual develops cancer. Since X and Y are binary, definition 2.4 of positive and negative causal relevance simplifies to the following: if $P(Y = 1 \mid do(X = 1)) > P(Y = 1 \mid do(X = 0))$, then X is a positive causal factor for Y; if the inequality is reversed, then X is a negative causal factor for Y; and if the two conditional probabilities are equal, X is not causally relevant to Y. The important point here is that when X and Y are binary, there are just three possibilities: X is positively relevant, negatively relevant, or irrelevant with regard to Y.

Suppose P is the human population of concern to the extrapolation. Thus, the question is whether X is a positive causal factor for Y in P, given that it is such among rats. Suppose we know that the mechanism set from X to Y is positively consonant. Then X is a positive causal factor for Y exactly if the relative frequency in P of those who possess an undisrupted mechanism is greater than zero. More formally:

> *Extrapolation Theorem*: Let X and Y be binary variables and let $\mathbf{M}_{XY}$ be the mechanism set from X to Y in P. Let φ_0 be the proportion of members of P for whom all mechanisms in $\mathbf{M}_{XY}$ are disrupted. Suppose that $\mathbf{M}_{XY}$ is positively consonant with respect to P. Then X is a positive causal factor for Y in P if and only if $\varphi_0 < 1$.

The proof of the extrapolation theorem is straightforward. If X is a positive causal factor for Y in P, then it is an immediate consequence of the disruption principle that $\varphi_0 < 1$. On the other hand, suppose that $\varphi_0 < 1$. Then there is a subset of P, call it Φ_a, such that $a > 0$ and $\varphi_a > 0$. From positive consonance, it follows that ΔP_a is not negative, and since $a > 0$ (i.e., Φ_a is a subset of individuals possessing a mechanism from X to Y), the disruption principle entails that ΔP_a is not equal to zero. Therefore, ΔP_a is strictly positive, and so is $\varphi_a \Delta P_a$. But since positive consonance entails that no $\varphi_i \Delta P_i$ is negative, it immediately follows from equation (6.9) that ΔP is strictly positive, that is, X is a positive causal factor for Y.

Comparative process tracing, as described in Chapter 5, would be the basis for the claim that there is a mechanism from X to Y in P, that is, for $\varphi_0 < 1$. Thus, the extrapolation theorem illustrates how the step from extrapolating a mechanism to extrapolating positive causal relevance can be made. However, the extrapolation theorem is limited not only insofar as it assumes consonance, but also in presuming that X and Y are binary. For instance, in the example of the substance found to be carcinogenic in laboratory rats, it is likely that X would represent quantity of dosage rather than exposure versus non-exposure. Let us begin with the question of what basis there might be for presuming consonance.

There is a circumstance in which positive consonance *must* be true, namely, if X is a necessary cause of Y. We can say that X is a necessary cause of Y just in case it is causally relevant to Y and $P(Y=0 \mid do\,(X=0)) = 1$. For instance, if you have not been exposed to HIV, then you certainly do not have AIDS. If X is a necessary cause, it is clear that positive consonance obtains, since a necessary cause can never be a negative causal factor. Besides the special case of necessary causes, there seem to be two general considerations that are relevant to assessing whether consonance is a reasonable assumption. A consistently null or positive impact in several populations in varied circumstances would support positive consonance. Knowledge of likely mechanisms through which X affects Y can also play an important role in assessing whether consonance is a reasonable assumption. Positive consonance may be a reasonable assumption when there is no plausible mechanism of any significance whereby the cause prevents the effect. Both of these motivations for consonance are present in the aflatoxin case, which may explain why researchers in this field seem to implicitly regard consonance as obvious. Of course, there are cases in which consonance is either highly uncertain or known to be false. Most apparently, if X is positively relevant to Y in some experiments and negatively relevant in others, there is clear evidence against consonance. Similarly, consonance is obviously not reasonable when it is likely that X is positively relevant to causes that exert opposite influences upon Y. In Appendix B, I examine how extrapolation of positive or negative causal relevance might be possible without assuming consonance.

Let us turn to a second limitation of the extrapolation theorem, namely, its restriction to binary variables. The assumption that X and Y are binary has the consequence that if X is causally relevant to Y, it is either positively or negatively relevant. As explained in section 2.3.2, the possible varieties of causal relevance are not so narrowly restricted when X and Y are quantitative variables. Recall the intuitive idea behind the general definition of causal relevance from Chapter 2 (definition 2.4): X is positively relevant to Y when increases in X yield increases in Y. Conversely, X is negatively relevant to Y when increases in X produce decreases in Y. When the relationship between X and Y is probabilistic, this intuitive idea requires some modification, since in that case, increases in a positive causal factor do not always result in increases in the effect. The most natural way to extend the proposal is to say that X is positively causally relevant to Y when increases in X yield increases in the expected value of Y.

If Y is not binary, then it is possible that X alters the probability distribution of Y without changing its expected value. In such a case, X would be causally relevant to Y while being neither positively nor negatively relevant. For example, a wealth redistribution program might change the distribution of income without changing its mean. Even setting aside such cases as these, when X is a quantitative variable, it may be causally relevant to Y without being either positively or negatively relevant. That point was illustrated by the fertilizer example (see Figure 2.7), in which the expected height of the plant increased with moderate dosages of fertilizer, reached a maximum, and then decreased with any further dosage elevation. In this case, the fertilizer is a positive causal factor for growth within moderate dosage intervals, and a negative causal factor in very high dosages. However, within an interval that spans both sides of the maximum, the fertilizer, though causally relevant to growth, is neither positively nor negatively relevant.

Thus, extending the extrapolation theorem to cases in which X or Y (or both) is a quantitative variable requires an additional premise. Given definition 2.4, one premise that would suffice is that $E(Y \mid do(x))$ is a monotonic function. A standard example of this occurs when the dependence of Y upon X satisfies the conditions of ordinary least-squares regression. In that case, Y is a linear function of X, the distribution of Y is normal, and the variance of Y is independent of X. Under these circumstances, if X is relevant to Y, it must be either positively or negatively relevant.

The fertilizer example illustrates that there are cases in which the average causal effect, $E(Y \mid do(x))$, is not monotonically increasing or decreasing for all values of X. In that example, the expected value of the height of the plant conditional on fertilizer increases monotonically with moderate dosages but decreases with higher ones. Thus, an application of the extrapolation theorem in this case requires strict attention to an interval of values of the candidate cause in which it may be reasonably presumed that $E(Y \mid do(x))$ is monotonically increasing or decreasing. For example, in the fertilizer example, we may be confident that within a moderate range of dosages the effect of fertilizer is monotonic. Clearly, disregarding the dosage interval in such a case could lead to mistaken extrapolations. For instance, the maximum point of the function $E(Y \mid do(x))$ might occur much earlier in some varieties or species than in others. In such a case, the fertilizer might promote growth for one plant variety within a given interval of dosages and inhibit growth within that same interval in another variety.

A consequence of the above analysis, then, is that careful attention must be paid to dosage levels in extrapolation when $E(Y \mid do(x))$ is nonmonotonic. One sometimes finds this same point expressed in discussions of animal extrapolation in the toxicology literature. For instance, a common theme in the literature concerning results of studies that aim to identify carcinogens in animal models for the purpose of making extrapolations to humans is that experimental animals are often exposed to far higher doses of the substance than humans would ever be likely to encounter. One article on this topic points out that:

> A linear dose response has been the dominant assumption in regulating carcinogens for many years, but this may not be correct. If the dose responses are not linear but are actually quadratic or hockey-stick shaped or show a threshold, then the actual hazard at low dose rates might be much less than the HERP [Human Exposure dose/Rodent Potency dose] values would suggest. (Ames et al. 1987, 272)[8]

The "dose response" is readily identified with the average causal effect, and the authors' point is that the reliability of an inference from high doses to low doses is very sensitive to the shape of this function. For example, if the function is linear, then the inference is unproblematic. However, the inference from high to low dosages is not reliable if the average causal effect is, say, "hockey-stick shaped," that is, is flat at low doses but sharply rising after a threshold is crossed. Some carcinogens, such as aflatoxin and the carcinogenic agent in tobacco smoke, appear to have linear dose response rates, while others, such as vinyl acetate, exhibit a well-marked threshold effect (Hengstler et al. 2003). Thus, it is an advantage of the present analysis of extrapolation that it implies that the interval of the cause under consideration matters when the average causal effect is nonmonotonic.

The aflatoxin example illustrates that the conditions specified by the extrapolation theorem are sometimes quite plausible in practice. As explained in Chapter 5, comparative process tracing provides good grounds for inferring that there is a mechanism from AFB_1 exposure to liver cancer among humans. Moreover, positive consonance is plausible in this case, because the carcinogenic effects of AFB_1 are consistently nonnegative (usually strictly positive) for a variety of animal models, and there appears to be no plausible mechanism whereby AFB_1 could prevent liver cancer. In addition, the effect of AFB_1 on liver cancer appears to be linear, which secures the key assumption of the extrapolation theorem that causally relevant factors are either positively or negatively relevant. Furthermore, the extrapolation theorem does not require that the model and target resemble one another in every causally relevant respect. There may be differences in one or more of the mechanisms linking cause and effect, as is indeed the case between Fischer rats and humans with regard to AFB_1. In fact, there may be some mechanisms present in one population that are completely absent in the other. Thus, the extrapolation theorem provides a more precise demonstration of the point that claims about positive or negative causal relevance can be extrapolated even when there are some causally relevant disanalogies between the model organism and the target.

Given this account of extrapolation of qualitative claims about positive or negative causal relevance, let us return to the subject of indefinite exclusive cp laws, which seem to play some role in science, yet for which no defensible interpretation has been suggested.

6.3 CETERIS PARIBUS AND EXTRAPOLATION

One reaction to the difficulties facing indefinite exclusive cp laws would be to say that the completer approach should be cast upon the junk pile of failed ideas and that to the extent that the expression "ceteris paribus" means anything, it must be interpreted according to one of the less problematic senses of "cp law." With regard to laws or empirical generalizations, I think that this is basically correct. But without further elaboration, this proposal fails to adequately address the role of "ceteris paribus" as it pertains to extrapolations. In the two subsequent subsections, I explain how this is so and how the analysis of extrapolation presented above helps to clarify this aspect of ceteris paribus.

6.3.1 Extrapolation in Extant Accounts of Ceteris Paribus

The relationship between cp laws and extrapolation is easily appreciated: "The law will hold, provided nothing interferes," is a hedged answer to the question of whether extrapolation is legitimate. The connection between ceteris paribus and extrapolation is noted by some philosophers. For example, Mitchell states that the challenge raised by cp laws is one of using knowledge about the circumstances in which the generalization has obtained to support inferences about whether it will also hold in a new situation (1997, S477; 2000, 256–57). Moreover, several accounts of cp laws explicitly connect the issue to extrapolation. But although there is merit to these proposals, none of them adequately address the type of extrapolation problem illustrated by the aflatoxin example. In that case, the generalization in question could not be transformed into a *definite* exclusive cp law in any nontrivial way, and it could not be presumed in advance that the generalization *usually* holds true with regard to the target populations of interest in the extrapolation.

Cartwright addresses the issue of cp laws in a way that draws attention to the extrapolation problem. According to Cartwright, cp laws are regularities derived from statements about capacities, which in her view are more fundamental. For instance, on her account, the law of universal gravitation asserts that massive objects have the capacity to attract one another with a force proportional to m_1m_2/r^2 (1999, 82–83). This statement about capacities is taken to entail the regularity that any pair of massive objects will attract one another with a force proportional to m_1m_2/r^2 in the ideal situation in which no force other than the mutual gravitation of the two objects is present. This is where "ceteris paribus" enters: "The regularities to be explained only hold *ceteris paribus*; they hold relative to the implementation and operation of a machine of an appropriate kind to give rise to them" (Cartwright 1995c, 279). Thus, the idea is that cp laws are statements about regularities that result from very particular and ideal arrangements wherein capacities can reveal their true natures.[9] This is a version of the definite exclusive interpretation of cp laws, but it is intended to have the virtue of explaining how such generalizations can be relevant even when the ideal conditions do not obtain, since the capacities endeavor to bring about their characteristic effects even when conditions are not ideal (Cartwright 1999, 82). According to Cartwright, then, cp laws are understood in terms of capacities, which in turn are said to underwrite extrapolation. Cartwright's capacity approach to extrapolation was critically examined in Section 5.2.

Glymour (2002) also approaches the issue of cp laws with an eye toward the problem of reliably extrapolating generalizations, but with formal learning theory[10] rather than capacities as his touchstone. Glymour sets up the problem as follows (2002, 400–401). We imagine a learner who makes conjectures about whether a certain generalization, call it X, will hold true in a sequence of instances. The learner may be an individual person or a group, such as a scientific community. The learner issues conditional conjectures of the form "If A, then X" or "If A, then not X." Let us call the first sort of sentence a *positive conjecture*, and the second sort a *negative conjecture*. Glymour proposes a criterion for whether a learner in such circumstances can effectively conjecture whether X will hold (2002, 401). The learner has *verified* "cp, X" if, in an infinite sequence of conjectures, the sum of false and negative conjectures is finite. Otherwise, the learner has *falsified* "cp, X."[11]

Thus, the learner verifies "Normally, X" or "cp, X" just in case there is a point in the infinite sequence after which she issues only true, positive conjectures. Hence, verifying a cp generalization on this proposal requires eventually developing some means for identifying conditions in which the generalization holds without exception. Another way to put Glymour's proposal, then, is that verifying a cp law is a matter of ultimately transforming an *indefinite* exclusive cp law into a *definite* one. Regarding an indefinite exclusive cp generalization as legitimate on this proposal would amount to conjecturing that this transformation will someday occur. This proposal has some plausibility with respect to examples from the history of physics. When such generalizations as the law of the pendulum, Boyle's law, and Galileo's law of free fall were originally proposed and used, it was known that they did not always obtain, yet the conditions in which they held without exception could not be specified in any exact way. As physical science progressed, it became possible to articulate such conditions on the basis of more fundamental theories. However, it is questionable whether comparable theoretical developments will occur in, say, cell biology and sociology, since it maybe that certain features of these fields make general unified theories very unlikely.[12] In any event, one would not want to make the legitimacy of generalizations in biology and social science contingent on the development of fundamental theories comparable to those in physics.

Finally, there is a straightforward connection between extrapolation and the interpretation of cp laws as normic generalizations. For if *normally* entails *usually*, then the proposition that, say, increases in the supply of labor normally produce decreases in wages supports extrapolations of this relationship to new cases. In fact, normic generalizations are quite closely related to the simple inductivist approach to extrapolation described in Chapter 5. Like normic generalizations, simple induction involves a default inference that is revocable under a set of somewhat open-ended conditions. But as was argued in section 5.1, extrapolation from animal models often cannot be justified on the basis of simple induction, a point illustrated by the aflatoxin example. For instance, it is not true that compounds carcinogenic at a particular site in rodents are usually also carcinogenic at that same site in humans. Moreover, the carcinogenic effect of aflatoxin B_1 varied considerably between distinct animal models, significantly promoting liver cancer in rats but having little or no effect in mice. In such a case, the reliability of the extrapolation cannot be settled by reference to generalizations asserting that what is true in a model organism is normally true of the target population. So, although normic generalizations are certainly an important part of extrapolation, they are far from being the whole story.

Existing proposals concerning cp laws, then, have addressed some aspects of the connection between extrapolation and ceteris paribus while leaving others relatively untouched. In particular, none of these proposals have explored the role of ceteris paribus in cases in which neither normic nor definite exclusive cp laws suffice for extrapolation. In the next subsection, I argue that an investigation of this territory reveals that the failings of the completer approach result from assuming that indefinite exclusive cp clauses qualify laws that are interpreted as universally quantified sentences. I show that the infirmities of the completer approach vanish if the indefinite exclusive sense of "ceteris paribus" is understood in relation to an inference schema that specifies sufficient conditions for extrapolating claims about positive or negative causal relevance.

6.3.2 Completers and Inference Schemas

Imagine a scientist who is the lead investigator of a recently published study showing that a particular compound is a cause of pancreatic cancer in rats. She is being interviewed by a science journalist who asks a question that will obviously occur to readers: Does this mean that the compound causes pancreatic cancer in humans, too? The scientist confronted with this question is likely to be in just the sort of situation carved out by Schurz's indefinite exclusive category. She cannot provide any nontrivial list of conditions in which it is always the case that the compound causes pancreatic cancer. Nor does she know that this causal generalization normally holds among mammals or whether it depends on special characteristics of rats. She has an idea of some factors that are likely to be important with regard to whether the extrapolation is correct, but she suspects that there may be others of which she is presently ignorant. In light of her own uncertainty and sensing that the journalist isn't really in the mood for a lecture on oncology, she responds, "Other things being equal, it would have the same effect."

Given the completer approach, we would say that the above utterance expresses the scientist's belief in a sentence of the form "Cp, all Fs are Gs," wherein "F" indicates exposure to the compound and "G," occurrence of pancreatic cancer; and "cp" means something like "as long as nothing interferes." The shortcomings of this proposal were discussed in section 6.1.2. In brief, when interpreted according to the completer approach, the scientist's claim ends up not being about the relationship between the compound and pancreatic cancer at all, but merely an assertion about the metaphysical doctrine of determinism. In other words, the completer approach would interpret the scientist as asserting a putative law that does not satisfy the domain specificity requirement.

Yet the scientist's claim cannot be plausibly interpreted according to any of the other types of cp law in Schurz's categorization either. The claim is not a comparative cp law. The statement that the compound causes pancreatic cancer in rats is indeed such a claim. But the statement that this effect would transfer to humans, "other things being equal," is not a claim *that* a particular causal relationship obtains, but rather a gesture at conditions under which it would. And the "other things being equal" is supposed to refer (albeit vaguely) to those conditions. Thus, the claim in this example is not merely a comparative cp law. And since the scientist cannot specify nontrivial conditions in which the causal relationship holds without exception, and does not know that it is typical among mammals, the claim is neither a definite exclusive nor a normic exclusive cp law.

At this point one might be inclined to say that the scientist's statement really is just a bit of insignificant fluff whose purpose is to brush off the reporter. But I think that this is too quick. After all, such expressions seem quite natural in the context of extrapolation problems like the one just described and in the aflatoxin example. Moreover, the problems that undermined the completer approach can be made to disappear if two modifications are made. First, take the generalization of concern to be a claim about positive causal relevance and not a universal generalization (our old friend "All Fs are Gs"). Second, interpret the cp clause as referring to an inference schema concerning extrapolation rather than to an empirical law.

"All Fs are Gs" is a singularly inappropriate format in which to represent the claim that exposure to the compound causes pancreatic cancer. The causal claim does not entail that everyone exposed to the compound develops pancreatic cancer, and this is presumably not the case. It does not even entail that there is some condition C such that it is a law of nature that everything that is both F and C is also G. The causal claim is consistent with the world being indeterministic or with there being no laws of nature concerning F, C, and G. The most natural way to interpret the claim that the compound causes pancreatic cancer is as an attribution of positive causal relevance. For instance, if exposure to the compound and the occurrence of pancreatic cancer are represented by the binary variables X and Y, respectively, then the scientist's research has shown that $P(Y = 1 \mid do(X = 1)) > P(Y = 1 \mid do(X = 0))$ among rats. That is, when exposure and non-exposure to the compound are determined by an ideal intervention, the exposure increases the chance of pancreatic cancer. If X and Y were quantitative variables, then positive causal relevance would mean (roughly) that interventions that increase the value of X also increase the expected value of Y.

It is not only more plausible to interpret the claim of interest to the extrapolation in terms of positive causal relevance rather than as a universal generalization. Doing so also eliminates the problem that the completer approach transforms generalizations qualified by a cp law into claims about determinism. Consider the difficulties involved in extrapolating a claim of the form "All Fs are Gs." A single F that is not G suffices to render this extrapolation incorrect. Furthermore, it assumed that no general theory is available that allows one to specify conditions in which the law holds without exception. There is no *definite* exclusive cp law waiting in the wings. In this situation, there is little more that can be said other than that the extrapolation will be correct unless there is something that causes an F not to be a G. Suppose, in contrast, that the generalization of interest is a claim concerning positive causal relevance. When the variables are binary, the extrapolation theorem (section 6.2.2) specifies conditions that suffice for the correctness of the extrapolation. And the conditions of the extrapolation theorem do not require the existence of deterministic causes. Thus, the argument that the completer approach makes cp laws equivalent to a claim about determinism depends upon the premise that the cp clause attaches to a universal generalization. If one supposes instead that the generalization is a claim about causal relevance, then the connection between determinism and indefinite exclusive cp clauses vanishes.

But simply replacing "All Fs are Gs" with a claim about positive causal relevance and leaving all other aspects of completer approach unchanged fails to resolve its fundamental shortcoming, namely, that it violates the domain specificity criterion. The domain specificity criterion stated that laws should provide information specifically relevant to their intended domain of application. Transforming cp laws into claims about the metaphysical doctrine of determinism is one way to violate the domain specificity criterion, but it is not the only way. For example, suppose that one interpreted the scientist's statement that the compound would cause pancreatic cancer among humans, other things being equal, as follows:

$$\text{Cp, } X \text{ is a positive causal factor for } Y, \tag{6.10}$$

where X and Y are binary variables indicating exposure to the compound and pancreatic cancer, respectively, and the "cp" states that positive consonance obtains and that some members of the target population possess an undisrupted mechanism from X to Y. There is no commitment to determinism lurking in (6.10), but interpreting cp laws in this manner would nevertheless contradict the domain specificity criterion. For (6.10) follows directly from the disruption principle; its truth is utterly independent of the relationship between the compound and pancreatic cancer. In short, (6.10) provides no information specifically about the subject matter of interest (the carcinogenic effects of the compound), but simply reflects a commitment to the disruption principle.

However, the conflict with the domain specificity criterion is avoided if the cp clause is understood in reference to an inference schema rather than to an empirical law or generalization. That is, suppose that, in the present context, "ceteris paribus" is interpreted as an all-purpose term for referring to conditions that would suffice for the extrapolation to be correct. In the case of extrapolating a claim asserting positive causal relevance, "ceteris paribus" could refer to the conditions articulated in the extrapolation theorem. On this way of understanding the matter, one has a claim about positive causal relevance made with respect to a base population and an abstract inference schema that indicates conditions that would suffice for the extrapolation of this claim to the target population. There is no violation of the domain specificity criterion here: claims about positive causal relevance provide domain-specific information about particular populations, while an inference schema is not an empirical law and need not be domain-specific.

An analogy with deductive logic may be helpful to convey the idea here. Suppose that Fred wishes to show that a particular hypothesis about the relationship between federal budget deficits and economic growth is false. For convenience, let us call this hypothesis H. Sue points out to Fred that he can disprove H if he can establish premises of the form "not-E and if H, then E." Thus, Sue is indicating a strategy for establishing the desired conclusion. However, it would be absurd to interpret Sue as suggesting that "If H entails E, and E is false, then H is false, too" is a cp law of economics. For this logical schema is a tautology, and hence says nothing specifically about economics at all. Tautologies may be laws of logic, but they are certainly not laws of economics, nor empirical laws of any kind. Similarly, it is a mistake to interpret (6.10) as a law about some scientific discipline: like *modus tollens*, it is not an empirical law but an inference schema. Whereas a law of economics must provide information specifically about economic phenomena, an inference schema may abstract entirely from the details of particular subjects. Of course, domain-specific information would be required to establish the premises needed to instantiate the inference schema in any given case, and providing this evidence is typically the hard part of extrapolation. The usefulness of an inference schema is that it indicates just what premises would suffice.

Wolfgang Spohn is the only writer I know who explicitly points out that some of the standard problems confronting cp laws evaporate if one treats cp clauses as something other than empirical generalizations.[13] Spohn writes:

> It is commonplace by now that laws or their applications are often to be qualified by some kind of *ceteris paribus* condition. As long as a law is conceived of as a proposition, the nature of this qualification is hard to understand. It seems to make the proposition indeterminate or trivial. But when we conceive of belief in a law as more than belief in a proposition, at least some of these mysteries dissolve in a quite natural way. (2002, 383–84)

Spohn develops an innovative version of the normic interpretation of cp laws that is rather different from my own account of extrapolation. But the above quotation is very much in the spirit of my diagnosis of the completer approach: a central failing of this proposal is that it interprets "ceteris paribus" as a qualification of a *law* in circumstances in which it indicates an *inference schema*, in particular, one specifying conditions that suffice for extrapolation. This is not to say, of course, that it is never appropriate to understand cp clauses as qualifications of laws. In terms of Schurz's categorization of cp laws, my claim concerns *indefinite exclusive* cp clauses that are *not* plausibly interpreted as saying *usually*, *typically*, or *normally*. This sort of situation is illustrated by the aflatoxin example and the imaginary example given at the head of this subsection. The completer approach fails in virtue of attempting to interpret such cp clauses as qualifications of laws construed as universally quantified generalizations. The difficulties confronting the completer approach go away if "ceteris paribus" is understood in reference to an inference schema indicating conditions that suffice for the extrapolation of a claim about causal relevance.

6.4 CONCLUSION

The expression "ceteris paribus" can be used to mean a remarkable variety of different things. Not only does the ambit of "cp law" encompass a diverse collection of generalizations, but cp clauses can also be used to qualify inferences, especially extrapolations. In this chapter, I have endeavored to show that considering ceteris paribus from the perspective of extrapolation sweeps away the failings of a common and quite problematic proposal on this topic, namely, the completer approach. The infirmities of the completer approach stem from two features. First, it interprets "ceteris paribus" as a qualification of laws in contexts where that expression refers to an inference schema that articulates sufficient conditions for extrapolation. Second, it assumes that the laws in question are universally quantified generalizations, archetypically of the form "All

Fs are Gs." The consequence of these two characteristics is that the completer approach transforms cp laws into claims about the existence of deterministic causes of the effect. Hence, existing versions of the completer approach violate the domain specificity criterion, according to which laws or other important empirical generalizations of a domain should provide information specifically about it. For instance, laws of economics should provide information specifically about economic phenomena. However, I showed that the failings of the completer approach can be avoided if indefinite exclusive cp clauses are understood as indicating an *inference schema* that specifies sufficient conditions for extrapolating claims about positive or negative *causal relevance*. Determinism is not an issue if the generalization in question is a probabilistic causal claim, and unlike an empirical law, an inference schema need not provide domain-specific information. Developing this proposal involved articulating sufficient conditions for extrapolating claims about positive and negative causal relevance, which was done in the extrapolation theorem. The extrapolation theorem rested upon the groundwork of the foregoing chapters, particularly the disruption principle. I explained how the extrapolation theorem applied to the aflatoxin example introduced in Chapter 5. Moreover, the extrapolation theorem reinforces the point that similarity in all causally relevant respects between model and target is not necessary for extrapolating claims about positive or negative causal relevance. In the next chapter, I turn to a philosophical issue that is intertwined with the mechanisms approach to extrapolation, namely, reductionism.

7

Reduction and Corrective Asymmetry

In the broadest terms, reductionism maintains that there are some levels of description of nature that are more fundamental than others. According to the reductionist, the more fundamental level of description and theorizing explains why generalizations at higher levels hold to the extent that they do, and accounts for their failures. In biology and social science, the fundamental realm would typically be conceived of in terms of part-whole relationships: descriptions of the basic components explain and correct generalizations concerning the behaviors of the systems constructed from them. These components would be macromolecules in the case of biology and individual agents in social science. Such a reductionist perspective fits snugly with the mechanisms approach to extrapolation. This proposal suggests that knowledge of underlying mechanisms and factors that interfere with them is especially valuable for specifying conditions in which a causal generalization will and will not obtain.

Yet reductionism is a highly controversial doctrine. The most common objection to it is known as the *multiple-realizability argument*. This argument rests on the premise that systems differing significantly with regard to basic causal mechanisms may nevertheless display some surprisingly similar behaviors. In such cases, the reasoning continues, an explanation given in terms of underlying mechanisms would miss important patterns displayed by an explanation that abstracts from those details. Thus, the multiple-realizability argument concludes, scientific explanation sometimes requires that details of basic mechanisms be omitted and that characteristics of systems be accounted for by way of higher-level descriptions. Such higher-level explanations are often said to be *autonomous* of the causal mechanisms formulated in terms of the basic components of the system. This train of thought is taken to support a position known as *pluralism*, according to which there is no level of description capable even in principle of achieving all scientific aims; rather, there are distinct forms of representation suitable for distinct purposes. If the mechanisms approach to extrapolation is linked with reductionism, then critiques of reductionism such as the multiple-realizability argument may be relevant to it. This leads to two questions. First, is the mechanisms approach to extrapolation indeed committed to reductionism? And second, if it is, do objections to reductionism undermine or at least substantially limit the applicability of the mechanisms approach?

Answering these two questions requires clarifying what is meant by "reductionism." I propose that reduction is an explanatory strategy that can be pursued in order to achieve a variety of goals, and what form the reduction should take depends on its purpose.[1] I argue, therefore, that there are no uniformly correct constraints on the form of reductions, since different types of reductive explanation might be suited to different objectives. I introduce four possible goals of reduction: ontological parsimony, unification, decomposition, and correction. I show how these four potential reductive goals can be used as the basis for a categorization of distinct varieties of reductionism and to specify which reductionisms are appropriately associated with the "reductionist anti-consensus."[2] Since this reductionist position is significantly weaker than some commonly critiqued versions of reductionism, I consider whether it has a legitimate claim to the title. My strategy is to connect reductionism to the notion of explanations drawn from a level of description that is more fundamental than others, to explicate the relevant sense of "fundamental" in terms of what I call *corrective asymmetry*, and to show that the reductionism in question satisfies this condition.

I propose that the motivation for the mechanisms approach rests upon the assumption that mechanisms are correctively asymmetric with regard to the claims of interest to the extrapolation. This proposition is the basis of the answers to the two questions posed above. First, the mechanisms approach to extrapolation is tied to reductionism insofar as it presumes the existence of mechanisms that are correctively asymmetric with regard to the generalizations to be extrapolated. Thus, the answer to the first question is a qualified *yes*: mechanisms-based extrapolation is committed to a form of reductionism. However, the answer to the second question—whether objections to reductionism threaten the mechanisms approach to extrapolation—is *no*. This is because the presence of correctively asymmetric mechanisms is entailed by a version of reductionism that is *not* undermined by the multiple-realizability argument.

An interesting consequence of this discussion is that the form of reductionism to which the mechanisms approach to extrapolation is linked is compatible with pluralism. There are three principles that I associate with this doctrine: that there are multiple legitimate strategies for representing nature (*principle of multiple perspectives*); that there is no ideal representation that is sufficient for all explanatory purposes (*non-completeness*); finally, that distinct levels of explanation are autonomous (*autonomy of levels*). I argue that the principle of multiple perspectives is consistent with even the most extreme version of reductionism, while non-completeness and autonomy of levels are consistent with the existence of correctively asymmetric mechanisms. Indeed, I suggest that corrective asymmetry is helpful for explicating the notion of autonomous levels.

7.1 ABSTRACTING FROM THE GORY DETAILS

Let us begin with a concrete example to motivate the multiple-realizability argument and pluralism. In the foregoing chapters, I have discussed how attention to mechanisms—often described in molecular terms—can aid in the refinement and extrapolation of such generalizations as "HIV causes AIDS" or "aflatoxin B_1 causes liver cancer." Such examples are surely grist for the reductionist's mill. But a critic of reductionism would be quick to point out that scientific understanding often requires representations that systematically ignore enormous amounts of gory detail concerning causal mechanisms.[3] That, in effect, is the point of the multiple-realizability argument. An illustration of this argument is provided by a continuation of the HIV example discussed in section 4.3.

As described there, M-tropic strains of HIV normally predominate in the early stages of HIV infection, while T-tropic strains become more prevalent in the later, symptomatic stage. The change in prevalence from M-tropic to T-tropic HIV is known as a "phenotype switch," and there is evidence that the switch is not merely a side effect but rather a contributing factor to immune failure and the onset of AIDS symptoms (Connor and Ho 1994; Glushakova et al. 1998). The phenotype switch is intimately related to the extent of resistance conferred by mutations that prevent the expression of the R5 co-receptor and thereby potentially block M-tropic HIV replication. In this context, Stine writes: "One of the great unsolved puzzles of HIV disease is why, during disease progression, does HIV lose its ability to infect macrophage and become T-cell tropic?" (2000, 140).

I found three hypotheses mentioned in the literature to account for why HIV infection almost always begins with M-tropic HIV (Zhu et al. 1993, 1180–81). The first, and least promising, is the *low inoculum* model. The essential idea here is something like the founder effect in evolutionary biology: an infection commences with an inoculum drawn at random, which is unlikely to exhibit much variation owing to its small size. The obvious difficulty with this proposal is that it fails to explain why the early and nonsymptomatic stages of HIV infection are invariably dominated by a *particular* type of HIV. At best, the low inoculum model can explain a trend from lesser to greater variability among the viral population present in a host as the infection progresses. But the hypothesis provides no explanation of why HIV in early stages of infection would nearly always be predominantly M-tropic.

A somewhat more promising hypothesis is the *selective transmission* model. According to this hypothesis, M-tropic HIV is more readily transmitted, through mucous membranes, for example, than T-tropic strains. One difficulty with this hypothesis is that the occurrence of the phenotype switch does not appear to depend on the mode of transmission, as one would expect if the selective transmission model were correct. For example, the same pattern of M-tropic-then-T-tropic prevalence has been found among hemophiliacs infected by HIV through direct blood transfusions (Zhu et al. 1993, 1180). In such cases, it is very doubtful that M-tropic HIV would be selectively transmitted vis-à-vis T-tropic HIV. Another puzzle for the selective transmission model is the long period in which M-tropic HIV predominates, followed by a rapid switch to T-tropic strains. The fact that the switch occurs, suggests that T-tropic strains can arise from M-tropic ones, which would hardly be surprising, given the high mutation rate of HIV. Thus, there is apparently something that keeps the proliferation of the T-tropic mutants in check until the later stages in the progression of the disease. But the selective transmission model seems incapable of specifying what this check might be or why it ultimately ceases to be effective.

Finally, there is the *selective amplification* model. According to this hypothesis, M-tropic HIV possesses a selective advantage in the earlier stages of infection, but—perhaps owing to the gradual exhaustion of the immune system—the fitness of T-tropic strains increases as the infection proceeds. The selective amplification model has the advantage of predicting a predominance of M-tropic HIV in the early and asymptomatic stages of infection, regardless of the mode of transmission. Moreover, Duncan Callaway, Ruy Ribeiro, and Martin Nowak (1999) have devised a mathematical model that shows how the selective amplification hypothesis can explain the phenotype shift. The most important premise in the model is that T-tropic HIV infects cells at a higher rate than M-tropic strains, but that the immune system mounts a more effect assault against the T-tropic variety.[4] Hence, the rough idea is that while M-tropic strains have a selective advantage at the start of the infection, their replication slowly weakens the immune system and ultimately tips the balance in favor of the more virulent T-tropic strains.

More specifically, let u_m and u_t be parameters representing the effectiveness of immune response to M-tropic and T-tropic strains, respectively, per unit of activated HIV specific T-helper cells. Thus, the overall effectiveness of the immune response to T-tropic strains depends upon both u_t and the quantity of T-helper cells that target HIV. Likewise, let β_m and β_t represent the replication rates of M-tropic and T-tropic strains. In the model, if $u_t/u_m > \beta_t/\beta_m$, then T-tropic HIV can proliferate only if the stock of HIV specific T-helper cells is sufficiently depleted by M-tropic infection (Callaway et al. 1999, 2525). As would be expected, the more u_t/u_m exceeds β_t/β_m, the longer the time before the switchover. The proportion of activated T-helper cells at the time of infection also affects the duration of M-tropic predominance. The larger the background of activated T-cells, the larger the target pool for HIV infection, and hence the more swiftly the T-helper cell stocks are depleted (Callaway et al. 1999, 2526–27).[5] For some values of these parameters, the switch from M-tropic to T-tropic strains never occurs, while for others, T-tropic strains predominate from the beginning. However, Callaway et al. report that the phenotype switch is a robust phenomenon that occurs in a broad range of the parameter space of their model (1999, 2527).

The selective amplification model, therefore, provides a plausible explanation of the phenotype switch. It is also a good example of the multiple realizability argument. According to the Callaway et al. model, the key factor implicated in the delay of T-tropic HIV proliferation is that $u_t/u_m > \beta_t/\beta_m$, and the key factor in the switchover is that u_t/u_m not exceed β_t/β_m beyond a critical point. Clearly, this situation is multiply realized. A wide range of parameterizations suffices for the phenotype switch, which is important, since it is obvious that immune response varies from one individual to the next and that replication rates vary among HIV strains. Moreover, much of the detail concerning the causal mechanisms of HIV replication is irrelevant to Callaway et al.'s explanation. The exact process by which HIV attaches to T-cells, by which the viral RNA is reverse transcribed, and so forth could change without altering the basic pattern of the phenotype switch so long as the crucial circumstances identified by Callaway et al. continued to obtain. Moreover, the model entails that the phenotype switch does not depend on the relative proportions of M-tropic and T-tropic strains at the very beginning of infection, as the low inoculum and selective transmission models propose (Callaway et al. 1999, 2526).

The Callaway et al. model also suggests fruitful hypotheses with regard to other puzzling features of HIV disease. For instance, there are several studies noting that the phenotype switch occurs less frequently among individuals infected with HIV C, a strain common in some parts of Africa and the Indian subcontinent (Abebe et al. 1999; Cecilia et al. 2000).[6] If the Callaway et al. model is correct, one would expect that u_t/u_m typically exceeds β_t/β_m by a greater amount among individuals infected with HIV C than among those infected with other HIV strains. The model is also suggestive with regard to the issue, discussed in section 4.3, of whether the homozygous thirty-two- base-pair deletion in the gene for the R5 co-receptor nullifies the effect of HIV exposure upon AIDS. Even if that mutation effectively blocks the replication of M-tropic HIV, it would not confer immunity if the values of the key parameters in the model were such as to allow proliferation of T-tropic HIV from the start of the infection. For example, this could occur in individuals whose immune systems did not mount an effective response to T-tropic strains. Since there is now at least one known case of an HIV-positive individual who is homozygous for the thirty-two-base-pair deletion in the gene for the R5 co-receptor (Biti et al. 1997), such a possibility seems worthy of exploration.

In sum, Callaway et al.'s explanation of the phenotype switch is an example of how advances in scientific understanding can ensue from abstracting from the nitty-gritty causal mechanical details. That of course is not to suggest that abstracting from such detail is always the best way to proceed. Rather, it is to say that the examination of complex systems in terms of the intricate details of the causal interactions of their components is not always the route to scientific discovery and explanation. With this motivating exemplar in hand, let us turn to an examination of reductionism, the doctrine that the multiple-realizability argument seeks to undermine.

7.2 WHAT'S REDUCTIONISM?

Whether the multiple-realizability argument is an effective objection to reductionism depends on just what that doctrine asserts. I characterized reductionism above as the thesis that there is some fundamental level of representation from which generalizations at higher levels can be explained and corrected. In the case of biology, the fundamental level would be what Sahotra Sarkar terms "macromolecular physics" (1998, 146–50). In the case of social science, the fundamental level would be interactions among individual persons. The multiple-realizability argument calls reductionism into question by pointing out that there are cases in which explanations given at less than fundamental levels are preferable. And this challenge to reductionism might also seem to cast doubt upon the mechanisms approach to extrapolation. However, I maintain that reductionism comes in several varieties, only some of which are subject to the multiple-realizability objection. And the version of reductionism that is not undermined by the multiple-realizability argument is all the reductionism that the mechanisms approach to extrapolation needs.

7.2.1 Four Motives and Three Desiderata

Reduction is an explanatory strategy that can, in different contexts, be pursued to achieve distinct goals. The four potential motives for reduction that I shall consider are the following:

> *Ontological Parsimony*: To show that where there appear to be two types of entity there is in fact only one, more fundamental type
>
> *Decomposition*: Given a feature possessed by a certain set of systems, to show that the systems' parts and their interactions, described at a specified level of detail, are sufficient to explain the feature
>
> *Unification*: To demonstrate that a wide array of distinct generalizations and observations can be explained by a small number of more fundamental generalizations
>
> *Correction*: To identify and explain exceptions to less fundamental generalizations, using generalizations and details drawn from a more fundamental realm.[7]

These motives are not intended to be exhaustive. Nor do I suggest that all of these four goals are always reasonable ones to pursue. In deciding whether a reductive research strategy is appropriate in a particular scientific context, one should ask whether the intended goal of the reduction is worthwhile and, if it is, whether reduction is an effective way to achieve it. Clearly, the answers to these questions can vary from case to case.

The classic example of a reduction that achieves the goal of ontological parsimony is the kinetic theory of heat, which is usually taken to show that heat is not a distinct substance, as Carnot's caloric theory maintained.

A stock example of a reduction that accomplishes unification is the explanation by Newtonian mechanics of a range of less fundamental laws such as Galileo's law of free fall and Kepler's laws of planetary motion. This case also illustrates correction, since Newtonian mechanics can be used to identify and explain exceptions to Kepler's laws. Reductions that aim to attain decomposition correspond closely to what Sarkar terms "strong reduction" (1998, 43–45). Sarkar defends the proposition that strong reductions are in the process of being carried out in molecular genetics (1998, chap. 6).

It will be helpful to examine the four reductive motives listed above along with conditions often demanded of reductions. Consider the following three:

1. The distinct concepts of the more and less fundamental realms must be linked by biconditional bridge laws or "synthetic identities."
2. Reduction must be a relation that holds among theories, where theories are understood as consisting of, or at least being associated with, a relatively small number of generalizations capable of accounting for a broad range of phenomena.[8]
3. Aside from correspondence rules linking the two realms, a reduction must explain the aspect of the less fundamental realm *solely* in terms of concepts and principles drawn from the fundamental one. For instance, a molecular explanation that presupposes various aspects of cellular context that are not described in molecular terms is not a reduction.

The model of reduction that anti-reductionists often take to be the standard account satisfies all of these conditions. This is what might be called the "layer-cake model."[9]

The classic presentation of the layer-cake model of reduction is found in Paul Oppenheim and Hilary Putnam's (1958) essay, "The Unity of Science as a Working Hypothesis." The layer-cake model presupposes that contemporary science justifies depicting nature in terms of a series of levels from most to least fundamental. In Oppenheim and Putnam's essay, these levels are elementary particles, atoms, molecules, cells, multicellular organisms, and social groups (1958, 9–10). The levels are intended to be such that the entities at any level above the fundamental one are fully decomposable into the entities at the level below. For example, molecules are decomposable into atoms, and atoms into elementary particles. It is presumed, in addition, that there is (or would eventually be) a set of theories for each level and that the terms in the theories at each level would refer exclusively to the entities at that level. Given this framework, the layer-cake model asserts that reduction consists of a deduction of the higher-level theory from the lower-level one with the aid of bridge laws that equate any distinct higher-level kinds with complexes of lower-level kinds. The layer-cake model, then, endorses all of 1 through 3; that is, according to it, reduction is a relation between theories, requires synthetic identities, and only resources drawn from the fundamental realm may be used in the derivation.

Although the layer-cake model has been called "the standard account of reduction" (Kincaid 1990, 576), it is rather different from that proposed by Ernest Nagel, whose account of reduction is often cited as the classic source on the topic.[10] Indeed, of requirements 1 through 3, Nagel's model insists only upon 2. Nagel did not require that the correspondence rules connecting the reducing to the reduced theory be synthetic identities or biconditionals; he is in general rather flexible as to what might qualify, and explicitly allows one-way conditionals (cf. Nagel 1979, 105–7). The insistence on synthetic identities was a modification of Nagel's model introduced by subsequent commentators, particularly Kenneth Schaffner (1967).[11]

Nagel's model also differs from the layer-cake model in not requiring that the more fundamental theory exclusively refer to entities that are proper parts of those referred to by the less fundamental theory. This is most evident in the case of what Nagel refers to as *homogeneous* reductions, that is, reductions in which the reduced theory contains no terms not already present in the reducing theory (1961, 342). Nagel's stock examples of homogeneous reductions are the derivations of Galileo's law of free fall and Kepler's laws of planetary motion from Newtonian mechanics (cf. 1979, 98). Moreover, there is nothing in Nagel's model to rule out the possibility that the reducing theory might span several levels of description, and thus violate requirement 3.[12]

Consider how the differences between Nagel's model and the layer-cake model regarding requirements 1 through 3 translate into differences with respect to the four reductive motives. A reduction that fit the form of the layer-cake model would succeed in achieving the first three of the motives listed above.[13] Synthetic identities would support the claim that any entity referred to by the reduced theory is identical to a complex of entities, or features of such complexes, described at the more fundamental level. Hence, such a reduction would achieve ontological parsimony, since it would show that no entity referred to by the higher-level theory constitutes an independent type of substance. Likewise, since the explanation proceeds solely from a more fundamental level of description, which is so characterized in virtue of referring to proper parts of the entities of the higher-level theory, the goal of decomposition would also be achieved. Since the layer-cake model assumes that reduction is a relationship between theories, understood as a reasonably small number of principles that encompass a broad range of phenomena, the goal of unification is met as well. In contrast, a reduction that fulfilled the strictures of Nagel's model would achieve the goal of unification—since what does the reducing is a theory—but not necessarily ontological parsimony or decomposition. For instance, these two motives are clearly irrelevant to the reduction of Kepler's laws of planetary motion to Newtonian mechanics.

In addition, a consideration of distinct motives for reduction illuminates the inherent implausibility of treating the layer-cake model as the *only* model of reduction.[14] Since not all attempts at reduction intend to achieve the goals of ontological parsimony, decomposition, and unification, it is unreasonable to insist that reductions in general must satisfy conditions relevant to the attainment of all of these goals. For instance, requirement 1, which demands that the terms of the reduced and reducing theories be connected by synthetic identities, is closely tied to the goal of ontological parsimony. Yet reductionists and anti-reductionists are united in their rejection of vitalism; the standard anti-reductionist position is *physicalist* anti-reductionism. Synthetic identities, furthermore, have little relevance to the other three reductive goals. Hence, there is no reason to suppose that reductions in molecular biology must in general have ontological parsimony as one of their goals, and consequently it is unreasonable to insist upon synthetic identities or biconditional bridge laws in biological discussions of reduction.[15]

7.2.2 Reductionisms

By the term "reductionism" I understand a substantive thesis about what sorts of reductions are possible in which areas of science. Thus, reductionism should be distinguished from models of reduction, which make claims about what characteristics an explanation must have in order to qualify as a reduction. A model of reduction might be judged correct or incorrect independently of any particular stance on reductionism. For example, the dispute between David Hull (1972; 1974) and Schaffner (1969; 1993b, 437–45) concerning the alleged reduction of Mendelian genetics to molecular genetics assumed Schaffner's (1967) model but turned on its application in this particular case. Given the four motives for reduction presented above, a variety of possible versions of reductionism can be distinguished. The most extreme is what can be called *hegemonic reductionism*, according to which the capacity of an explanation given at a higher level to achieve any of the four goals is equaled, or surpassed, by an explanation provided at a more fundamental level. Although Oppenheim and Putnam's classic (1958) paper is plausibly interpreted as an endorsement of hegemonic reductionism, it is doubtful that the position has any current philosophical defenders. An example of a more restrained version of reductionism would be the claim that the unifying power of any higher-level explanation can be matched or exceeded by an explanation at a more fundamental level. This position is naturally labeled *unifying reductionism*. Unlike hegemonic reductionism, unifying reductionism does not require that the more fundamental theory describe entities that are proper parts of those referred to by the theory to be reduced.

The standard objection to reductionism, the multiple realizability argument, provides a good basis for rejecting hegemonic and unifying reductionism. Although this argument can be expressed in various ways (cf. Fodor 1975; Kitcher 1984, 1999; Rosenberg 1985, 93–96), I propose that it be interpreted as endeavoring to show that molecular explanations are sometimes (perhaps often) less unified than explanations provided at a higher level. Putting the matter in somewhat different language, the central claim is that heterogeneous collections of molecular mechanisms sometimes underlie what are, from a higher-level perspective, single generalizations. Hence, an explanation that replaced the higher-level generalization with its underlying molecular detail would suffer a loss of unifying power. Putnam's classic exposition of the multiple-realizability argument maintains that a geometrical explanation of why a round peg 1 inch in diameter won't fit into a 1-inch diagonal square hole is superior to one couched in terms of molecular structure in virtue of being more general (1975, 296). Robert Batterman (2000, 2002) proposes that multiple realizability be interpreted with regard to the physical concept of "universality," which concerns physical systems that exhibit similar patterns of behavior in spite of being constituted of distinct materials. For example, the law of the pendulum holds (approximately) of pendulums whether they are constructed of iron, copper, or plastic. In Batterman's formulation, universal phenomena are characterized by the following two features:

1. The details of the system (those details that would feature in a complete causal-mechanical explanation of the system's behavior) are largely irrelevant for describing the behavior of interest.
2. Many different systems with completely different "micro" details will exhibit the identical behavior. (2002, 13)

So, just as with Putnam's example of the round peg and the square hole, although an explanation at the level of microdetail might be possible in principle, such an explanation would miss the common pattern seen in the simple geometrical explanation. Likewise, attempting to explain the phenotype switch at the level of molecular interactions would obscure the key importance of the ratios u_t/u_m and β_t/β_m in accounting for the phenomenon. The multiple-realizability argument, then, is aptly summed up in Harold Kincaid's statement "Attempts to explain in purely biochemical terms will tend to see diversity where there is important unity" (1990, 587).

The multiple-realizability argument has attracted a great deal of criticism. One line of objection is that anti-reductionists have exaggerated the extent to which multiple realizability is a genuine problem, and thereby have overstated the heterogeneity of molecular explanations in biology. For example, Joseph Robinson (1992, 465) takes this line of argument in his response to Kincaid (1990). The response is also pursued by Sarkar with respect to the relationship between molecular and Mendelian genetics (1998, 159–68), while William Bechtel and Jennifer Mundale (1999) make an analogous argument for neuroscience and psychology. For instance, Bechtel and Mundale argue that despite obvious variations in neurological details within and across species, there are nevertheless important similarities in neurological structure that play a fundamental role in explaining such things as visual perception. The theme of this line of objection, therefore, is that multiple realizability is not a challenge to reduction so long as it is possible to identify relevant commonalities at the level of the reducing theory (cf. Hooker 2004, 442–43, 470–75). Differences in molecular detail are consistent with similarities that can be characterized in molecular terms and are capable of accounting for general patterns. For example, in the case of the phenotypic switch, it is presumably not a coincidence that T-tropic strains tend to replicate at a more rapid rate, while the immune system mounts a less effective response to M-tropic strains. It seems likely that there are general molecular features of these two strains of HIV that account for the difference.

I think that the sources cited in the foregoing paragraph make a good case for a doctrine that one might call *mitigated unifying reductionism*: molecular explanations of higher-level biological phenomena are often, though not necessarily always, unified. However, the objection does not show that multiple realizability fails to pose a genuine objection to unifying (and hence hegemonic) reductionism. Unifying reductionism entails that the unification attainable from any higher-level explanation can be equaled or surpassed by an explanation given at a more fundamental level. In contrast, mitigated unifying reductionism does not assert this. Most obviously, it allows that there are cases in which there is no unified lower-level explanation corresponding to a higher-level one. A more interesting point, however, is that even if there is a unified explanation at a fundamental level, the higher-level explanation may nevertheless be simpler, have greater scope, and be more efficient. In a word, the higher-level explanation may be more unified. That would contradict unifying reductionism, which requires not only that the explanation at the fundamental level be unified, but also that it be *at least as unified* as any other explanation. In Putnam's example of the round peg and the square hole, it seems quite doubtful that an explanation in terms of the common molecular features of wooden, plastic, and metal pegs and boards punched with holes could match the geometrical explanation in the small number of generalizations required to account for the phenomena. A similar point seems plausible with respect to the explanation of the phenotype switch provided in the foregoing section.

A second line of criticism of the multiple-realizability argument is that reduction, and explanation in general, can serve purposes other than unification. I take this to be Kenneth Waters's point when he asserts "The unificationist criterion for explanation is implausible when invoked within the nitty-gritty details of genetics" (1990, 136). More recently, Elliott Sober has argued in a similar vein (1999, 549–51), maintaining that while the lower-level account may not be unified, it nevertheless explains. Sober asserts, moreover, that it is "a matter of taste" as to whether the unified but shallow or the heterogeneous but deep explanation is preferable (1999, 550–51). Sober also argues that there are sometimes sound reasons for preferring an explanation that includes molecular detail at the expense of unity, especially if this detail makes it possible to correct the higher-level generalization (1999, 555–56).

The point of these arguments is that one can have reduction without unification, and that there is sometimes good reason to desire such reductions. The type of reduction in question can be characterized as follows: for any feature of a complex system, there is a reduction of that feature which achieves decomposition, though not necessarily unification. This is similar to what Jerry Fodor calls "token-token reductionism." Token-token reductionism asserts that in each particular case, the parts and their interactions can, in principle, explain the features of the whole. One important reason why token-token reductionism is of interest is, as Sober observes, that it entails that knowledge of the underlying details will allow for corrections to higher-level generalizations. This point brings up a further species of reductionism, namely, *corrective reductionism,* which states that the resources of the more fundamental level are always capable of correcting higher-level generalizations. Token-token reductionism entails corrective reductionism, since if every instance can be explained by way of components and their interactions, then any exception to any higher-level generalization can also be thus explained. The entailment does not go in the opposite direction, however, since correction might be had without decomposition—as the example of Newtonian mechanics and Kepler's laws of planetary motion illustrates.

In sum, the reductionist anti-consensus can be interpreted as maintaining the conjunction of mitigated unifying reductionism and token-token reductionism, along with the observation that token-token reductionism entails corrective reductionism. But it might be objected that the only reductionism that should count as such is hegemonic reductionism. For example, according to Kincaid, the only legitimate interpretation of reduction is one that asserts that:

> One theory reduces another when it can do all the explanatory work of the reduced theory.... If there were good reasons to think that molecular biology and statistical mechanics could not do all the explanatory work of their higher-level counterparts (and there is), then whatever they have achieved, it is not reduction. To claim reduction while admitting explanatory incompleteness is to make the issue a trivial semantic one. It is not. (1997, 5)

Although Kincaid is right that it would be pointless to defend reductionism merely through a redefinition of terms, it is equally the case that one ought not to criticize reductionism by knocking down a straw man. As we saw above, the layer-cake model of reduction, which Kincaid assumes as the standard account (1990, 576; 1997, 50), is more stringent than Nagel's model, has few if any current defenders, and is implausible considered on its own merits. Once one distinguishes between causal-mechanical explanations that aim to elucidate underlying processes from theoretical explanations designed to achieve unification, it is plausible that a reduction might accomplish one of these explanatory purposes but not the other.

But if one can be a reductionist without advocating hegemonic reductionism, it is fair to ask what distinguishes reduction from other sorts of explanatory relationships that might hold between distinct levels of inquiry. Surely, examinations of a phenomenon from diverse perspectives can be mutually illuminating in a variety of ways, but not all of this can count as reduction if the term "reduction" is to mean anything. On what basis, then, can a nonhegemonic reductionism lay a genuine claim to the title?

7.2.3 Corrective Asymmetry

Reduction rests on the idea that some levels of explanation are more fundamental than others; for example, hegemonic reductionism asserts that there is a level of explanation that is most fundamental in virtue of being able to equal or surpass any goal attainable by any explanation provided at any other level. Clearly, one who rejects hegemonic reductionism but nevertheless claims to defend a reductionism of some sort requires a different interpretation of "fundamental." I propose that the key notion here is *corrective asymmetry*. Roughly put, corrective asymmetry means that resources from the fundamental level are necessary to correct explanations provided at other levels, *but not vice versa.*

A bit more needs to be said about the term "level" in order to make this rough statement of corrective asymmetry more precise.[16] I shall assume that levels are distinguished on the basis of the *resources* associated with them. These resources include concepts and entities posited, together with generalizations and other statements formulated in terms of these concepts and entities. For instance, the resources of molecular biology would include such concepts as hydrogen bonding, various important macromolecules such as DNA, and such generalizations as the standard account of protein synthesis. In contrast, classical genetics would invoke distinct concepts (e.g., gene, dominance, etc.), would not mention the characteristic entities of molecular biology, and would rely upon distinct generalizations. When the entities of one level constitute the parts from which the entities of a second level are composed, then the second level is said to be *higher* than the first. Thus, Mendelian genetics is a higher level than molecular genetics. However, the difference between levels need not correspond to a part-whole relationship. For example, consider the relationship between Newtonian mechanics and what might be called phenomenological planetary astronomy, which would prominently include Kepler's laws of planetary motion. Newtonian mechanics contains concepts (e.g., gravitational force, mass), and refers to entities (e.g., absolute space) and generalizations (e.g., the law of universal gravitation) not found in phenomenological planetary astronomy. But the relationship of the entities referred to at the two levels is not one of part to whole. It is likely that there is often some overlap between levels and, moreover, that the exact boundaries of levels are somewhat vague. Nevertheless, I think that there are frequently cases in which tolerably clear distinctions can be drawn between levels, as the above examples illustrate.

Consider two levels, which we may for convenience label L_1 and L_2. I will say that L_1 is *correctively asymmetric* with respect to L_2 if and only if the resources of L_1 can correct L_2 in some situations in which L_2's own resources would not suffice for this purpose, but the reverse is never the case. If L_1 is correctively asymmetric with regard to L_2, then I shall judge L_1 to be *more fundamental* than L_2. For example, Newtonian mechanics explains many exceptions to Kepler's laws of planetary motion, such as those that arise from the perturbing gravitational force of a second planet, yet Kepler's laws do not explain failures of Newtonian mechanics. *That* function is performed by a more fundamental theory, namely, general relativity.

If token-token reductionism is true, then molecular biology is correctively asymmetric with respect to higher levels of biological description. In the case of HIV replication, for example, the molecular details correct higher-level descriptions of the process. As we saw, the ability of molecular biology to correct and refine generalizations stated at higher levels, even when no unified molecular explanation is in the offing, was the basis of one criticism of the multiple-realizability argument. Of course, this does not mean that corrections can *never* be made on the basis of anything other than molecular biology. Rather, the claim is that it is at least sometimes, and probably often, the case that the correction can be had no other way. In contrast, token-token reductionism entails that resources drawn from higher levels are never necessary to correct molecular explanations of particular biological events. Notice that this does not mean that such corrections could never be stated in higher-level terms; rather, the claim is that any correction stated in such terms could be replaced by one drawing solely upon molecular resources. Let us consider this more carefully.

Consider how one might explain why a certain strain of HIV is resistant to a particular anti-retroviral drug, for example, a class known as non-nucleoside reverse transcriptase inhibitors (cf. Stine 2000, 85–87). These drugs interfere with HIV replication by binding to the enzyme reverse transcriptase, which catalyzes the reverse transcription of the viral RNA to viral DNA that is then incorporated into the DNA of the host cell. By binding to reverse transcriptase, non-nucleoside reverse transcriptase inhibitors prevent it from carrying out its normal function. However, there are strains of HIV that are resistant to such drugs as a result of possessing mutations that alter the molecular structure of reverse transcriptase, and in some cases the relevant changes in the base pairs of the viral RNA are known. It is clear that there is a straightforward selective explanation of the prevalence of such mutant strains in individuals treated with non-nucleoside reverse transcriptase inhibitors. Does this constitute a counterexample to corrective asymmetry?

It is not difficult to argue that the answer to this question is no. Consider the biological events of which one would claim a molecular explanation in this case, for instance, how a particular change in the viral genome results in the synthesis of a distinct version of reverse transcriptase. The evolutionary explanation described above does not claim to correct such explanations, since, if token-token reductionism is true, then they depend solely on the molecular facts of the case. The same is true of particular cases of infection or failures of infection of individual cells by particular HIV. The prevalence of one strain of HIV over another in an HIV+ person results from the summation of a large number of such events, each molecularly explicable. Of course, the explanation that appeals to natural selection is more unified than the one that provides the gory molecular details of each case, but that does not conflict with token-token reductionism or corrective asymmetry.

Another concern is that molecular explanations presume a context that is characterized in higher-level (e.g., cytological) terms. For example, a description of HIV replication presupposes the existence of cells of several types and their organelles, as well as the larger organ systems (e.g., lymphatic) in which they occur. But variations in these contextual features might account for exceptions to the usual molecular processes. However, there is a simple response to this concern. If token-token reductionism is true, then the relevant contextual features can be described in molecular terms in each individual case; hence, the higher-level description is not necessary to explain the exception.[17]

Corrective asymmetry is also exhibited by reductions that achieve explanatory unification. Newtonian mechanics explains exceptions to Kepler's laws of planetary motion, but Kepler's laws do not return the favor. Thus, corrective asymmetry explicates a shared sense of "fundamental" operative in reductions that achieve unification and those that accomplish decomposition. In both the Newtonian and the HIV replication examples, the fundamental level is the one that has, as it were, the last word about what happens. Consequently, using corrective asymmetry as a criterion for what distinguishes reduction from other sorts of explanatory relationships between distinct levels has the appealing feature of being able to account for how genuine reductions may come in several forms. Notice that such a plurality of forms of reductive explanations does not fit comfortably with hegemonic reductionism.

Hence, an advocate of token-token reductionism would qualify as a reductionist if corrective asymmetry is taken as a mark of what makes one level more fundamental than another. However, the same cannot be said of mitigated unifying reductionism. The fact that some molecular explanations achieve a significant measure of unification does not entail that there is a corrective asymmetry between molecular biology and other levels of biological description. But that is not to say that mitigated unifying reductionism is unimportant or of no interest. Mitigated unifying reductionism blunts the effect of the multiple-realizability argument. Furthermore, mitigated unifying reductionism is important for the mechanisms approach to extrapolation. Recall that mechanisms, as defined in section 3.4.1, exhibit regular patterns of behavior, a point illustrated by the HIV replication mechanism described in chapter 4. Thus, mechanisms-based extrapolation in biology is founded not only on the premise that the molecular level is correctively asymmetric with regard to higher levels. It also presumes that processes at the molecular level can be characterized as mechanisms, which requires that regular patterns of behavior be discernible in these molecular processes. Fortunately, there is good reason to think that this is indeed the case and that claims of "wildly disjunctive" molecular processes are wild exaggerations.

In the subsequent section, I argue that the conjunction of token-token and mitigated unifying reductionism is consistent with pluralism. Moreover, I endeavor to show that mitigated unifying reductionism and corrective asymmetry are important for clarifying and defending the pluralistic doctrine of autonomy of levels.

7.3 CAN A REDUCTIONIST BE A PLURALIST?

At the end of his classic statement of the anti-reductionist position in biology, Philip Kitcher wrote, "Despite the immense value of the molecular biology that Watson and Crick launched in 1953, molecular studies cannot cannibalize the rest of biology" (1984, 373). Kitcher's use of the word "cannibalize" gives an indication of the consequences that anti-reductionists fear would ensue were reductionism correct. Apparently, if reductionism were right, then all provinces of biology other than molecular biology would be, in principle at least, superfluous—of practical use only because of limitations in computing power and knowledge of initial conditions. To one strongly attached to these allegedly superfluous disciplines, this would be a dire consequence indeed. Not surprisingly, then, philosophers who advocate pluralism—roughly, the thesis that there is a plurality of legitimate and autonomous levels of description and explanation of a given phenomena—often see themselves as staking out a position that stands in direct opposition to reductionism. In this section, I argue that pluralism is consistent with reductionism and corrective asymmetry, and that the latter concept helps to clarify the notion that higher levels of may be autonomous.

7.3.1 Core Principles of Pluralism

Several authors have defended pluralism (cf. Dupré 1993; Cartwright 1999; Longino 2000, 2002a, 2002b; Kitcher 2001; Mitchell 2002b, 2003), and although there are some differences of detail and emphasis among them, I think that the following three principles are a good characterization of their core position.

Principle of Multiple Perspectives: There are multiple legitimate strategies for representing nature.

Non-Completeness: There is no ideal representation that is sufficient for all explanatory purposes.

Autonomy of Levels: Distinct levels of explanation are autonomous.

The first two of these principles are succinctly articulated by Kitcher, who writes:

> The pluralism I propose consists of the following claims: (1) there are many different systems of representation for scientific use in understanding nature; (2) there is no coherent ideal of a complete account of nature....[18] (2002, 570)

Clearly, (1) is a statement of the principle of multiple perspectives, while (2) is non-completeness. John Dupré encapsulates pluralism as follows:

> The most general positive doctrine I shall advocate is pluralism: first, in opposition to an essentialist doctrine of natural kinds, pluralism as the claim that there are many equally legitimate ways of dividing the world into kinds, a doctrine I refer to as "promiscuous realism"; and second, in opposition to reductionism, pluralism as the insistence on the equal reality and causal efficacy of objects both large and small. (1993, 6–7)

Dupré's "promiscuous realism" is a version of the principle of multiple perspectives, while I interpret the second of his two claims as a statement of autonomy of levels.

There are, I believe, two primary motivations for pluralism: one pragmatic and the other ontological. The pragmatic motivation rests upon the sensible notion that, owing to differences in goals and interests, one phenomenon can be legitimately studied from a variety of perspectives. Given that there is no one objectively correct set of goals and interests to have (cf. Kitcher 2001, chaps. 4–6), it follows that there is no one objectively correct perspective from which to pursue one's inquiries and that it is doubtful that a complete representation of the world is possible. A "complete representation" in the sense at issue in non-completeness would be one that would suffice (in principle) for the achievement of any purpose that any other representation could accomplish. Kitcher supports non-completeness via an analogy with maps (2001, 55–63). Maps are representations designed to serve certain purposes, and although there may be a map that is sufficient for a given set of aims, it is implausible that there could be a map of (say) the Earth that would suffice for all goals that an earthbound traveler might conceivably have (2001, 60). By analogy, scientific representations are devised for particular ends, so we should be likewise skeptical that there is, even as an ideal, a single representation that would serve all ends. The ontological motivation for pluralism appeals to features of the world, especially, its complexity. If the world were very simple, it might be feasible to devise a single representation of it that could answer any question one might ask, but since the actual world is staggeringly complex, no such ideal representation is possible. For example, according to Mitchell, ''The complexity of nature and the idealized character of our causal models to explain that complexity conspire to entail an integrated pluralistic picture of scientific practice'' (2002b, 67). The idealized character of representations is presumably related to the complexity of the phenomena (it is not possible to include all relevant factors) as well as pragmatic concerns (one might be interested in only one aspect of the phenomenon).

So, can one consistently be both a reductionist and a pluralist? The answer depends on the type of reductionism one has in mind. Let us begin with hegemonic reductionism; clearly, this ought to be inconsistent with pluralism. And it is easy to see that it conflicts with non-completeness, the proposition that there is no one complete representation of nature. For if hegemonic reduction is correct, then any legitimate explanatory purpose that one wishes to achieve can (in principle) be attained via the most fundamental level. Hence, the representation provided at the fundamental level would constitute the complete account whose existence (and coherence) is denied by non-completeness.

Notice, however, that hegemonic reductionism is consistent with the principle of multiple perspectives. The hegemonist could happily agree that, for obvious practical reasons, various representational strategies are expedient, and hence legitimate. Consequently, the hegemonic reductionist can accept that there are many distinct, legitimate ways of conceptualizing nature. Of course, hegemonic reductionism maintains that there is one most fundamental level of description, but there is no apparent reason why only the most fundamental description should be regarded as legitimate. In sum, hegemonic reductionism is inconsistent with pluralism, but not in virtue of conflicting with the principle of multiple perspectives. Since hegemonic reductionism is logically stronger than both mitigated unifying and token-token reductionism, this is a useful conclusion, for if P entails R, and P is consistent with Q, then R is also consistent with Q. As a result, both mitigated unifying and token-token reductionism are consistent with the principle of multiple perspectives.

The issue, then, is whether these two versions of reductionism are consistent with non-completeness and autonomy of levels. There is a straightforward argument that they are consistent with non-completeness. Token-token reductionism asserts, in effect, that it is always possible (in principle) to provide a causal explanation of a particular biological event at the molecular level. This does imply that representations at this level are sufficient for a particular set of explanatory purposes, namely, causal explanations of particular biological events and such things (e.g., correction) that might ensue from these explanations. However, it does not entail that all explanatory aims can be thus achieved. Likewise, mitigated unifying reductionism claims only that some molecular explanations are unified. This is consistent with some not being unified or being less so than explanations given at other levels. For example, an explanation of the prevalence of a drug-resistant strain of HIV provided in terms of natural selection is more unified than one provided exclusively in terms of the molecular details of the replication, or failure of replication, of each particular virus. Hence, mitigated unifying and token-token reductionism are compatible with non-completeness. To return to Kitcher's "many maps" analogy, the claim that there is no map that suffices for all purposes does not rule out the possibility that there is a map that suffices for some particular purpose.

7.3.2 Autonomy and Unification

Let us turn, then, to the autonomy of levels. In order to decide whether this principle is consistent with mitigated unifying and token-token reductionism, it will be necessary to clarify just what "autonomy" amounts to in this context. Fodor defines autonomy as follows: "I will say that a law or theory that figures in bona fide empirical explanations, but that is not reducible to a law or theory of physics, is ipso facto *autonomous*" (1997, 149). Thus, autonomy involves two things: not being reducible and serving as a basis for "bona fide empirical explanations." Let us consider these two aspects of autonomy more carefully.

On the face of it, it seems easy to argue that autonomy of levels is consistent with token-token and mitigated unifying reductionism. That is, one need only interpret the "reduction" in the definition of autonomy as reduction that aims to achieve unification. This interpretation is reasonable, given that the primary motivation for the autonomy of levels is the multiple-realizability argument, which attempts to show that higher-level explanations are sometimes more unified than those formulated at lower levels. But this conclusion is consistent with both token-token and mitigated unifying reductionism. Rejecting token-token reductionism, in contrast, requires maintaining that there are strongly emergent properties, that is, properties of a whole that cannot be explained by its parts and their interactions *even on a case-by-case basis*. There are well-known puzzles associated with the proposition that there are strongly emergent properties,[19] and at least one pluralist, namely Kitcher, is clearly uncomfortable with such metaphysical excrescences (cf. 2002, 571). Likewise, rejecting mitigated unifying reductionism would require making the implausible claim that explanations given at a fundamental level are *never* or *very rarely* unified. In contrast, rejecting unifying reductionism only requires maintaining that lower-level explanations are sometimes less unified than higher-level ones. So, by interpreting the claim that there is no level to which all others are reducible to be a claim about unifying reductionism, a pluralist can agree with the reductionist anti-consensus.

But this reconciliation of reductionism and pluralism is illusory if an influential argument due to Jaegwon Kim (1992) is sound. Kim argues that if unifying reductionism fails, then higher-level generalizations cannot serve as a basis for genuine scientific explanations. In other words, Kim's claim is that if "reduction" means unifying reductionism, then the two defining features of autonomy are mutually incompatible. Kim's argument begins with the premise that a generalization can underwrite scientific explanations only if it is *projectible,* that is to say, positive instances of the generalization provide evidence for further positive instances in the future (1992, 11). From here the argument proceeds as follows (cf. 1992, 15, 18–20). Suppose that unifying reductionism fails as a result of multiple realizability, as the pluralist claims. Then higher-level expressions correspond not to a single physical kind but to a heterogeneous collection of them. Yet if this is so, there is no reason to expect that characteristics of one instance of the multiply realized higher-level kind will recur in future instances. But that is just to say that higher-level generalizations are not projectible after all, and hence, from the initial premise, not a potential basis for genuine scientific explanation.

Kim's argument does indeed point out a problem for some versions of the multiple-realizability argument that (as noted in section 7.2.2) tend to exaggerate the "wildly disjunctive" nature of physical realizations of biological and psychological phenomena. Moreover, Kim surely is correct that heterogeneity poses challenges for projecting generalizations; that is what the problem of extrapolation in heterogeneous populations is all about. Nevertheless, I think that Kim's argument is unsound. The problem centers on just what the conclusion of the multiple-realizability argument is. Consider these two possible interpretations:

1. Unifying reductionism is false.
2. Mitigated unifying reductionism is false.

Kim's argument would be quite reasonable if multiple realizability were understood according to (2). Suppose that (2) was the conclusion of the multiple-realizability argument. Then explanations from the more fundamental level would nearly always be extremely heterogeneous, and the sort of "wild disjunction" imagined by some advocates of multiple realizability would be the general rule. In this situation, Kim's argument seems quite compelling. But things are otherwise if (1) and not (2) is the conclusion of the multiple-realizability argument.

The Callaway et al. model described in section 7.1 is an example of how (1) is consistent with projectible generalizations. In that model, the phenotype switch was shown to be largely independent of all but a few crucial features, in particular, the relative replication rates and effectiveness of immune response to M-tropic and T-tropic HIV strains. In this example, the higher-level explanation of the phenotype switch is simpler and more efficient, and makes one less likely to lose sight of the forest for the trees. Thus it is reasonable in this case to say that the explanation provided by the Calloway et al. model is more unified than one given in molecular terms. Nevertheless, mitigated unifying reductionism is reasonable with regard to HIV research, wherein there are known molecular mechanisms of some generality. Thus, the phenotype switch appears to be an example in which the truth of mitigated unifying reductionism suffices for projectible higher-level generalizations despite the failure of unifying reductionism.

The above diagnosis of Kim's argument is reinforced by the discussion of extrapolation from chapters 5 and 6. Following Nelson Goodman (1954), Kim and his commentators think of projectibility in terms of universally quantified conditional sentences of the form "All Fs are Gs." In this situation, projectibility means that an F that is also G provides support for the expectation that subsequent Fs are G as well. But their examples are almost exclusively claims about positive or negative causal relevance; for instance, "Pains cause anxiety reactions" (Kim 1992, 16) or that ibuprofen ameliorates rheumatoid arthritis symptoms (Block 1997, 113). As explained in section 6.3.2, "All Fs are Gs," is an unsuitable format for representing claims concerning positive causal relevance. But an F that is also G is not properly regarded as an "instance" of a claim concerning positive causal relevance, since the individual might be both F and G solely by coincidence. The most natural way to interpret projectibility with respect to claims of positive or negative causal relevance is in terms of extrapolation. One learns (e.g., by randomized controlled experiment) that X is a positive causal factor for Y in one population and infers that it is also such in another. More specifically, a causal claim is *projectible* with regard to a set of populations if and only if learning that the causal claim is true of one population in the set provides evidence that it also true of the others.

Chapters 5 and 6 explored how mechanisms might play a role in justifying extrapolations. Extrapolation on the basis of mechanisms is pertinent to Kim's argument that multiple realization undermines projectibility, since extrapolation on this basis would be precluded if the mechanisms in the two populations were totally dissimilar. Nevertheless, mechanism-based extrapolation can proceed even in cases in which there are causally relevant differences in mechanism between the populations in question (see sections 5.4.2 and 6.2.2). In particular, how similar model and target must be depends upon the specificity of the causal claim to be extrapolated. This point was illustrated by the aflatoxin example in which differences between rat and human suggested that the effect of AFB_1 on liver cancer in rats is less than that in humans. Thus, although it would be unwise to extrapolate the exact causal effect from rat to human in this case, extrapolating positive causal relevance is quite reasonable. In short, neither extrapolation nor projectibility requires perfect homogeneity with regard to mechanisms.

7.3.3 Autonomy and Causal Reality

The upshot of the foregoing discussion is that by rejecting unifying reductionism, a person who accepts mitigated unifying and token-token reductionism can also agree to the autonomy of levels. However, this reconciliation of autonomy and reductionism might seem to be overly dependent on inevitably subjective judgments of relative unifying power. Autonomy is often thought to involve an attribution of real causal powers to the autonomous level, not merely to express a pragmatic preference for unified explanations. Dupré's statement, cited above, that pluralism entails "the equal reality and causal efficacy of objects both large and small," is naturally interpreted as an expression of this sentiment.[20] The equality of causal efficacy is apparently intended as an objective feature of the world, and not just a reflection of human interests. On the other hand, it is unclear that a more robust autonomy of this sort can be countenanced without postulating strongly emergent properties and thereby abandoning physicalism (cf. Rosenberg 1997, 2001). In this section, I argue that there is a very sensible way to understand what it is for properties at higher levels to be causally real that is consistent with token-token and mitigated unifying reductionism.

One straightforward way to interpret what it is for a property to be "causally efficacious" or to have "real causal powers" is the following. The property p is *causally real* just in case there are accurate generalizations asserting that p is a cause of something.[21] But higher-level properties could be causally real in this sense even if hegemonic reductionism were true. For if p is equivalent to some more fundamental property that is related to other properties by causal laws, then p must be causally real in the sense just given. Thus, a pluralist like Dupré might regard this sense of "causally real" as insufficiently robust. What more might "causally real" mean, then?

Let us approach this question in terms of what is typically treated as the "fundamental" level for the purposes of discussions of reduction in biology, namely, what Sarkar terms "macromolecular physics."[22] Note that macromolecular physics is not quantum mechanics or even chemistry, as Sarkar makes clear (1998, 146–50). For instance, macromolecular physics generally ignores the subatomic structure of atoms, often treating atoms as solid spheres, an assumption which, although usually good enough for the purposes of molecular biology, is clearly not accurate according to modern chemistry, much less quantum mechanics. What, then, justifies styling macromolecular physics a "fundamental" level of description? My answer to this question was provided above in section 7.2.3. Molecular biology is fundamental to the extent that it is correctively asymmetric with regard to other levels of biological description, such as classical genetics. But macromolecular physics is obviously *not* correctively asymmetric with regard to chemistry or quantum mechanics. One would suppose that the reverse is the case.

But in spite of the fact that molecular biology is hardly fundamental physics, it is unreasonable to deny that properties described at this level are causally real. The justification for the causal reality of molecular properties seems to consist in two things: there are accurate causal generalizations that can be stated at this level, and this level is correctively asymmetric with regard to some other levels. Yet these two features can be possessed by many other levels as well, such as evolutionary biology or microeconomics. Thus, if accurate causal generalizations and corrective asymmetry make properties characterized at the level of macromolecular physics causally real, why shouldn't the same go for other levels, too?

Consider how all of this connects to the mechanisms approach to extrapolation. The underlying premise of this approach is that knowledge of mechanisms and of factors that interfere with them is a guide for the correction and extrapolation of positive causal relevance and other probabilistic causal claims. The motivation for this premise is that the mechanism is characterized at a level that is correctively asymmetric with regard to the claims of interest to the extrapolation. But the requisite corrective asymmetry might be attained at any one of several levels, and there may often be sound scientific reasons for not delving more deeply than necessary. Chief among these reasons are the following: (1) the details required for the fundamental explanation are unknown; (2) even if these details were known, the fundamental explanation would be computationally intractable; (3) the fundamental explanation, even if it could be carried through, would obscure significant patterns exhibited in the higher-level explanation. All three of these considerations are relevant in motivating macromolecular physics rather than quantum mechanics as a basic level of biological explanation. Such considerations can also justify characterizing some biological mechanisms at a level higher than macromolecular physics. For instance, the Callaway et al. model is correctively asymmetric with regard to a description of the phenotype switch. Moreover, both (1) and (3) can be invoked in this case. Not all of the molecular details are known, and the exclusively molecular explanation might obscure the important causal pattern displayed in the Callaway et al. model.

Yet some levels will have a more extensive range of corrective asymmetry than others. For instance, molecular biology is correctively asymmetric with regard to a broader range of generalizations than the "level" consisting solely of the resources employed by the Callaway et al. model. There is an important practical implication of this simple observation. If one wants to know at what level to seek a mechanism for a particular correction or extrapolation, it is reasonable to choose a level that one is confident is correctively asymmetric with respect to the generalization of concern. In biology, this translates into a general justification for emphasizing the search for molecular mechanisms. Nevertheless, a general prescription of this kind is compatible with often describing mechanisms at levels other than the molecular one, for the reasons given above. In sum, the mechanisms approach to extrapolation is linked to reductionism through its connection to corrective asymmetry, but it is pluralistic insofar as corrective asymmetry may be had at several distinct levels. And there may be good scientific reasons for preferring mechanisms characterized at some level higher than that of fundamental physics, macromolecular physics, or, in social science, higher than the level of interactions among individual agents.

No doubt, some pluralists would be unsatisfied with this brand of pluralism. There are certainly *some* pluralist perspectives that are utterly incompatible with reductionism of any sort.[23] However, that is consistent with there being reasonable interpretations of reductionism and pluralism according to which both may be true.

7.4 CONCLUSION

Debate between defenders and critics of reductionism is a perennial theme in the philosophy of biology. Moreover, this topic is intimately linked to the mechanisms approach to extrapolation. The privileged role attributed to mechanisms by this approach depends upon their being correctively asymmetric with respect to the causal claims of interest to the extrapolation. Corrective asymmetry in turn can be used to explicate the concept of "fundamental level" inherent in reductionism, and to do so in a way that identifies a version of reductionism that is consistent with the multiple-realizability argument and with pluralism. Corrective asymmetry can often be had at more than one level of description, and there are often sound scientific reasons—including those emphasized by the multiple-realizability argument—for not descending more deeply than necessary.

The discussion so far has tended to focus on biological examples, with social science receiving little attention. That is reversed in the subsequent two chapters, which explore the prospects of utilizing the mechanisms approach to extrapolation in social science.

8

Extrapolation in Social Science

The foregoing chapters have examined the mechanisms approach to extrapolation mainly from the perspective of examples drawn from biology. This chapter and the next address the question of whether mechanisms-based extrapolation can be usefully employed in social science. I begin with an examination of the only extensive methodological examination of extrapolation that I know of in the philosophy of social science, namely, Francesco Guala's book *The Methodology of Experimental Economics* (2005). Although I think that there is much of value in Guala's discussion, I argue that his account fails to address the basic challenges to extrapolation described in Chapter 5. Specifically, his account provides no answer to the extrapolator's circle and no explanation of how extrapolation can be possible even when there are causally relevant differences between the model and the target. Since the proposals advanced in Chapters 5 and 6 were intended to address these challenges, I examine the applicability to social science of the mechanisms approach to extrapolation described there.

In this chapter, I discuss two challenges to this methodological transfer: first, that the contemplated intervention might be likely to alter the relevant mechanisms, and second, that there may be a great deal of uncertainty about what the mechanisms are. These two issues are illustrated by a pair of case studies. One premise of the mechanisms approach to extrapolation, discussed in Chapter 3, is that mechanisms can be identified with causal structure. Making the case for this identification requires defending the claim that mechanisms provide information about how probability distributions change under interventions. But the possibility that policy interventions will restructure social mechanisms is a commonly posed challenge for social science. I provide an explication of the concept of a structure-altering intervention, and explore the circumstances under which an intervention is more likely to be structure-altering and how such changes can be anticipated.

This discussion leads directly to the first case study, which concerns extrapolation from experiments designed to evaluate the effectiveness of welfare-to-work programs. A central methodological disagreement concerning the evaluation of welfare-to-work programs in fact turned on the usefulness of social mechanisms for extrapolation. I argue that there were indeed good reasons to be skeptical of the prospects of a thoroughgoing mechanisms approach to extrapolation in this case, one of the most important of which is that the mechanisms, even if they could be accurately ascertained, would likely be altered by the proposed policy intervention. Nevertheless, insofar as they provided some qualitative indication of "reasons to suppose otherwise," attention to mechanisms was important for a conscientious application of simple induction. The welfare reform example, then, is a case in which mechanisms are useful for extrapolation, but not in the thoroughgoing way of the aflatoxin example described in Chapters 5 and 6. The second case study derives from experimental economics and examines the extrapolation to real-world settings of a now well-established result concerning "preference reversals." In this case the chief obstacle to extrapolation is not structure-altering interventions, but rather uncertainty about which mechanism explains the result. I describe two possible mechanisms and explain how they lead to very different conclusions about the prevalence of preference reversals outside the laboratory walls.

I see no reason in principle that mechanisms-based extrapolation cannot be successfully utilized in social science. Nevertheless, these case studies show that the extent to which this is in fact possible depends on how stable social mechanisms are under interventions and to what extent accurate knowledge of these mechanisms is attainable, which are matters that cannot be settled by philosophy alone.

8.1 GUALA ON EXTERNAL VALIDITY

Since the mid-1950s, experimental economics has enjoyed an enormous growth in research output and, correspondingly, in the collection of established results (cf. Roth 1995; Guala 2005). From the beginning, extrapolation—often referred to as "external validity"—has been a central methodological problem for experimental economics. In fact, until relatively recently, mainstream economists often dismissed the results of economic experiments as irrelevant to the behavior of real-world markets (Guala 2005, 2–3). Experimental economics, therefore, is a rich source of social science examples in which extrapolation is a genuine issue. Yet there is surprisingly little in the way of a methodological analysis of extrapolation issues with regard to this field. As Guala puts it:

> To write on external validity is challenging. Philosophers of science, surprisingly, have very little to say about it. Experimental economists also tend to ignore or downplay the relevance of external validity; they typically say that it is not a particularly useful concept and, moreover, that worrying too much about it may turn attention away from more important issues of experimental design. (2005, 142)

Consequently, an examination of Guala's proposal is a good way to begin this chapter. I argue that although much of what Guala says about extrapolation in experimental economics is very sensible, his account fails to answer the basic challenges to extrapolation described in Chapter 5. In particular, his account does not answer the extrapolator's circle, nor does it explain how extrapolation might be possible even when there are causally relevant differences between the model and the target.

Guala provides several detailed case studies of extrapolation from economic experiments. One of these examples is the case of auctions of broadcasting licenses conducted by the Federal Communications Commission (FCC) in 1993–94 that were designed with the aid of experimental economics (2005, chap. 8). The task was to design and implement, on short notice, an auction mechanism that would satisfy several desiderata specified by the FCC (2005, 162–63). One of these desiderata was the generation of revenue for the FCC, which meant attempting to award licenses to those willing to pay the most for them. Due to several complexities inherent in the broadcast license auction, the optimal auction mechanism could not be inferred from existing theory (2005, 166). Consequently, researchers proceeded by evaluating a small number of proposed mechanisms in the laboratory, in an effort to find which one worked best and to identify likely sources of difficulty in attempting to implement a particular mechanism (2005, 170–78). One mechanism was selected and implemented in the real auction in October 1994, with results that conformed to expectations based upon experiments, and which were generally regarded as successful (2005, 179–81).

Generalizing from this and other examples, Guala makes several suggestions about the conditions under which extrapolation from experiment to the real world is justified. For example, one of his main themes is that improving the internal validity of an experiment often makes it less similar to the real-world systems it is intended to model (2005, 144). Guala also notes that the claim that the experimental model is relevantly similar to the target is an empirical hypothesis (2005, 195). Concerning the nature of this empirical hypothesis and the means by which it is to be tested, Guala writes the following:

> In this case, the evidence is the correspondence between observed features of the target and observed features of the experimental system; the external validity hypothesis is that the relata belong to similar causal mechanisms. (2005, 197)

This is an endorsement of a mechanisms approach to extrapolation. However, as explained in Chapter 5 (see section 5.3), an invocation of mechanisms does not suffice to answer the extrapolator's circle. Extrapolation is motivated by ethical or practical limitations on directly studying the target system. Yet comparing mechanisms presumably requires studying both the model and target populations separately, and it is unclear how that can be done, given the limitations on what can be learned about the target by studying it directly. One of the chief goals of the proposal concerning comparative process tracing presented in section 5.3 was to answer this challenge. The central theme of comparative process tracing is that extrapolating a mechanism from model to target depends upon antecedent knowledge of stages of the mechanism at which significant differences are and are not likely to arise. Moreover, it is often possible to restrict attention to downstream stages of the mechanism upon which upstream differences must leave their mark. Thus, comparative process tracing explains how limited, partial information about the target can establish the model as a basis for extrapolation.

Although Guala does not explicitly describe comparative process tracing, one might suppose that this was what he intended. For example, the design of the FCC broadcasting license auction is arguably a case of comparative process tracing, since one of the experimenters' aims was to identify "fragile" points in mechanisms at which things were like to go awry (2005, 173). Thus, special attention would have to be paid to those fragile points when implementing the real auction.[1] But whatever his intentions, there are some aspects of Guala's proposal that are not compatible with the account of comparative process tracing given in Chapter 5. In particular, Guala endorses LaFollette and Shanks's criterion for a causal analogue model (CAM). Recall that in LaFollette and Shanks' terminology, a CAM is a model that can serve as basis for extrapolation to a specified target. According to LaFollette and Shanks, a model is a CAM only if there are no causally relevant disanalogies between it and the intended target of the extrapolation (see section 5.4.2). Guala cites LaFollette and Shanks approvingly (2005, 195) and reiterates their criterion of CAM-hood in more than one place (cf. 2005, xi, 199). For example, Guala writes, "The inference from experiment to real world is a special kind of analogical argument, in which the inference is strengthened by making sure that the two systems are similar in all relevant (causal) respects" (2005, xi).

LaFollette and Shanks use their strict criterion of CAM-hood to argue that animal models can never serve as a basis for extrapolation to humans, since there is always bound to be some causally relevant disanology. As pointed out in section 5.4.1, LaFollette and Shanks's criterion, if accepted, would not only rule out extrapolation from animal models to humans but also extrapolation from one human group to another. After all, there are bound to be some causally relevant disanalogies between any pair of human groups one might choose. Thus, one inclined to be the least bit optimistic about extrapolation in biology and social science cannot consistently accept LaFollette and Shanks's criterion of CAM-hood. Fortunately, we saw that there is a good reason to reject their criterion, namely, that it overlooks the connection between the specificity of the causal claim to be extrapolated and the criterion of similarity required of the model. Thus, extrapolating a claim about positive or negative causal relevance does not require a model that is similar in all causally relevant respects (see sections 5.4.1 and 6.2.2). In the aflatoxin example, for instance, the relevant mechanism in the Fischer rat differs from that in humans in way suggesting that the carcinogenic impact of AFB_1 is less severe in rats than in humans. This difference in mechanisms is surely a causally relevant disanalogy, but it does not suggest that extrapolating the positive causal relevance of AFB_1 to liver cancer from rats to humans would be a mistake.

At some points, Guala appears to retreat from LaFollette and Shanks's strict CAM criterion. For example, he writes, "The trick is to make sure that the target and the experimental system are similar in *most* relevant respects so as to be able to generalize from the laboratory to the outside world" (2005, 217; my italics). Guala does not comment upon the switch from "all" to "most," nor does he explain how some causally relevant disanalogies might be unproblematic. Answering this question requires distinguishing those causally relevant disanalogies that indicate it would be an error to extrapolate a particular type of causal generalization. That is the distinction illustrated by the aflatoxin example: there is a causally relevant disanology, but it does not suggest that it would be an error to extrapolate a claim that exposure to AFB_1 is a positive causal factor for liver cancer. The key point is that similarity in all causally relevant respects is not a necessary criterion for extrapolating qualitative causal claims, such as claims about positive or negative causal relevance.

Causally relevant differences are inevitable between and within populations studied by biologists and social scientists. Thus, a fundamental challenge for any account of extrapolation in these fields is to explain how extrapolation can be possible even in the presence of causally relevant disanalogies. Although there is much wisdom and good sense in Guala's account of extrapolation, his proposal has failed to meet this challenge and has not adequately responded to the extrapolator's circle. In contrast, the approach to extrapolation advanced in this book is intended to address those challenges. So, let us consider the prospects of extrapolation in social science from the perspective of the account developed here.

8.2 ARE SOCIAL MECHANISMS CAUSAL STRUCTURE?

The underlying premises of mechanisms-based extrapolation were traced in Chapters 2 through 6. One fundamental premise is the disruption principle: a causal effect is nullified just in case every mechanism from cause to effect is severed. Since a nullified causal effect means that ideal interventions on the cause make no difference to the probability of the effect, the disruption principle links mechanisms to probabilities. That connection was mediated by the identification of mechanisms and causal structure discussed in Chapter 3. But Chapter 3 left the basis for the identification of social mechanisms and causal structure unresolved. While a general, default argument for this identification was suggested for mechanisms in molecular biology on the basis of evolutionary theory, the prospects of an analogous argument in social science are, at least for the present, uncertain. I take up the thread of this discussion again in the following two subsections.

8.2.1 Structure-Altering Interventions

A standard challenge to social science focuses on the ability of human beings to alter their own social organization (cf. Taylor 1971; Fay 1983; Searle 1984, chap. 5). Nagel concisely expressed the concern as follows:

> A third difficulty confronting the social sciences, sometimes cited as the gravest one they face, has its source in the fact that human beings frequently modify their habitual modes of social behavior as a consequence of acquiring fresh knowledge about the events in which they are participating or the society of which they are members. (1961, 466)

In economics, the concern that interventions might alter the institutional structures or social practices upon which predictions of the policy's effects are based is associated with the "Lucas critique" (Lucas 1981). Robert Lucas argued that econometric models capable of making accurate short-term economic forecasts would often fail to correctly predict the consequences of policy interventions, since the intervention would change "the decision rules of agents" (1981, 110–11). Since social mechanisms are patterned complexes of interactions among agents that generate macrosociological regularities (see section 3.5.1), such changes amount to alterations in social mechanisms.

The potential for social mechanisms to change in response to interventions challenges their identification with causal structure, and thereby the extension of mechanisms-based extrapolation to social science. Recall that causal structure is that which generates probability distributions and provides information regarding how those probability distributions change under interventions. As discussed in section 3.4.2, *modularity* is an important feature of causal structure. Modularity requires that interventions at a given point in the structure leave downstream causal relationships unaltered. The question, then, is whether social mechanisms are modular in this sense.

I shall call interventions that violate modularity *structure-altering*. Structure-altering interventions are best understood in contrast to ideal interventions. Recall that an ideal intervention is an exogenous direct cause of exactly one variable in a system, eliminates all other causal influences that would ordinarily affect the variable it targets, but directly changes nothing else (see item (a) in definition 2.1). For example, suppose that the graph in Figure 8.1 represents the causal relationships among participation in an abstinence-only sex education program (P), attitudes about sex (A), and having children out of wedlock (W). An ideal intervention on the variable P is represented in Figure 8.2. An ideal intervention erases all arrows pointing into P but otherwise leaves the structure unchanged. For example, in Figure 8.2, the causal relationship between A and W is exactly as before.

A structure-altering intervention, in contrast, makes changes in the causal relationships besides blocking the usual influences upon the

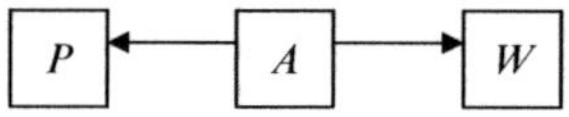

Figure 8.1 A common cause

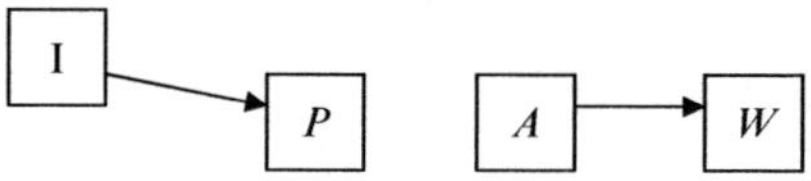

Figure 8.2 An ideal intervention

targeted variable. The concept can be defined more precisely as follows. Suppose the variable X is a member of a set of variables $\mathbf{V}$ that represent features of the system of interest. An intervention on X is *structure-altering with respect to* $\mathbf{V}$ just in case it changes causal relationships among the variables of $\mathbf{V}$ in addition to eliminating the causes of X. In Figure 8.2, $\mathbf{V} = \{P, A, W\}$ and the intervention is on P. The intervention in this case would be structure-altering if, for example, it modified the relationship between A and W. Notice that an intervention that directly affects more than one variable is not necessarily structure-altering, since targeting more than one variable need not entail altering the causal relationships among them.

It is also important to distinguish structure-altering interventions from the difficulty of differentiating causation from mere correlation. As Figure 8.2 illustrates, a correlation between two variables might result from the presence of a common cause of both, rather than from one being a cause of the other. Thus, the correlation between P and W in Figure 8.2 is a *mere* correlation: it is not invariant under *any* ideal intervention that targets P. One of the primary challenges to causal inference from statistical data is distinguishing between correlations due to unmeasured common causes and those resulting from a direct causal relationship. However, that problem is distinct from the possibility that a causal relationship might breakdown under certain types of intervention. For instance, the ideal gas law is invariant under some ideal interventions, and hence is a causal generalization. Nevertheless, the ideal gas law breaks down under interventions that set the pressure to a very high value. In this case, the fragility of the generalization is due to its rough and approximate character. But even a generalization that precisely describes a causal relationship in a given context might be subject to structure-altering interventions. Causal relationships typically depend upon background conditions too numerous and complex to fully and explicitly incorporate into a model. Consequently, interventions that change such background conditions may be structure-altering. Even if the generalization accurately described the causal relationship under the original set of background conditions, it might be an inaccurate representation of that relationship in the new circumstances brought about by the intervention.

Consider James Scott's account of the effects on the social structure of a Malaysian village of a government-sponsored irrigation project that made it possible to grow two rice crops per year rather than just one (1985, 74–85). Initially, the project significantly improved the economic situation of villagers, from the land-poor who relied on wage labor to the larger landowners. The increase in the supply of rice sharply reduced the threat of famine and, initially, the additional rice production doubled the demand for field labor, thereby significantly raising the income of poorer villagers. However, landowners soon discovered that renting combine-harvester machines was better than hiring field labor under the new system, since double-cropping required quickly harvesting one crop so that the next could be planted. Not only did this turn of events adversely affect the wages of land-poor villagers, it also undermined traditional demonstrations of generosity through which wealthy farmers attempted to ensure reliable sources of labor in the future. These practices included sumptuous feasts to which all in the village were invited, bonuses for laborers at the end of the harvest, and tenancy agreements that made allowances for poor harvests. In short, the innovation of double-cropping fundamentally altered the economic structure of mutual dependence between poor and wealthier villagers and all of the practices that went along with it.

Let R indicate the annual rice production, D the demand for field labor, and E the wage earnings of villagers. Finally, let I represent the intervention, that is, the government-sponsored double-cropping program. This example can be represented by the graphs in Figure 8.3. The intervention on R is structure-altering with respect to $\{R, D, E\}$. The intervention significantly attenuated the influence of rice production upon demand for field labor, which is represented in 8.3(B) by the deletion of the arrow from R to D. An arrow directly from the intervention to D is included in the graph in 8.3(A) because the intervention affected the demand for labor through a path not passing through R. That is, growing two crops per year rather than one placed a higher premium on a quick crop harvest, thereby increasing the use of combine-harvester machines and undermining the demand for field labor. Notice that the causal relationship in 8.3(A) would be invariant under many interventions on R that did not increase the number of crops per year. But the graph in 8.3(A) implicitly treats the harvesting of one crop per year as a stable background condition, and

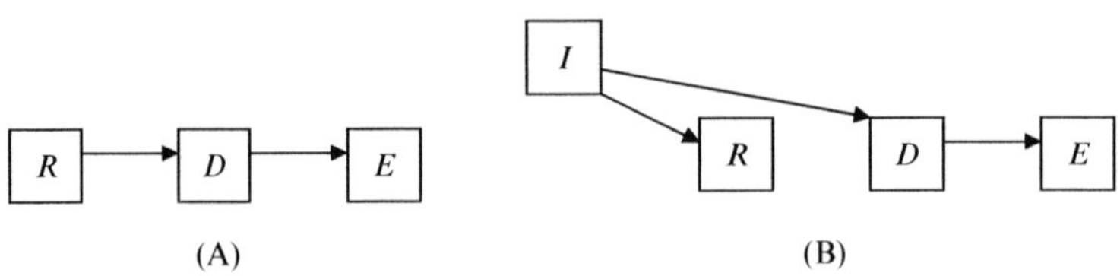

Figure 8.3 A structure-altering intervention

hence no longer correctly represents the influence of rice production upon the demand for field labor when that background condition is changed.[2]

8.2.2 Anticipating Changes in Mechanisms

The possibility that a policy intervention may be structure-altering, then, is far from an idle concern. Let us consider the circumstances relevant to whether structure-altering interventions are a serious concern and what might be done when they are. Whether a given intervention is structure-altering depends in part on the nature of the intervention itself. For instance, in the Malaysian example described in the foregoing section, an intervention that consisted solely of making fertilizer more readily available to rice cultivators would have been unlikely to significantly alter the structure represented in Figure 8.3(A). Furthermore, whether an intervention is structure-altering may depend upon the level of detail at which the mechanisms are represented. Recall that an intervention is said to be structure-altering with regard to a set of variables used to represent a causal system. Consequently, a single intervention may be structure-altering with regard to one set of variables but not another. Thus, although double-cropping was structure-altering with regard to the set of variables {*R*, *D*, *E*}, it might not have been structure-altering with regard to a finer-grained causal model that represents the motivations and decisions of individual cultivators. In general, some types of interventions are more likely to be structure-altering than others, and whether a given intervention is structure-altering may depend upon the variables chosen to represent the causal system. Let us consider these two points in turn.

Smaller-scale interventions are less likely to be structure-altering than large scale-ones. The category of "large-scale" intervention is rather vague, but examples would include such things as basic reforms of important organizations and the implementation of some major new social program. The government-sponsored irrigation projects and double-cropping in Malaysia, described in section 8.1.1, would thus qualify as a large-scale intervention. Large-scale interventions are contrasted with interventions that simply tweak some preexisting feature within an established structure. Examples of such smaller-scale interventions include a 10 cent increase in liquor tax or providing funding for a city to hire twenty new police officers. The scale of an intervention is not the only consideration pertinent to its structure-altering potential. Another relevant feature is the extent to which it is new or unprecedented. Interventions of a familiar type—say, the Federal Reserve raising the federal funds rate by a quarter percent—are less likely to generate new social structures for the simple reason that such reactive practices are already well entrenched. Lucas notes that our ability to predict changes in the decision rules of agents resulting from an intervention depends on whether the intervention follows some familiar, established pattern (1981, 119–20). In sum, large-scale interventions without recent precedents have the greatest potential to be structure-altering. Although this criterion is vague, it nevertheless can distinguish some cases, for example, a large-scale irrigation scheme versus a quarter-percent hike in interest rates. The welfare reform example, discussed in section 8.3, also illustrates that this rough criterion can be useful.

Suppose, then, that one is confronted with a contemplated intervention that is likely to be structure-altering with regard to a mechanism at a specific level of description. In this situation, the obvious thing to do is to attempt to anticipate, on the basis of some more fundamental theory, the potential changes in causal structure likely to be wrought by the intervention (cf. Nagel 1961, 471; Lucas 1981, 124–26). In the case of social phenomena, that more fundamental theory might be one that focuses on individual interactions. The most fully developed theory of this sort is based on the notion of rational choice, in the sense of maximizing expected utility: changes in the mechanism can be anticipated by considering how rational actors would respond to the incentives created by the new policy.[3] This suggestion appears quite plausible in the Malaysian example described above. Rational choice theory could predict that landowners would switch to combine-harvester machines in the double-cropping regime that placed a higher premium on a quick harvest.

I think that there is some merit to the suggestion that changes to social mechanisms resulting from policy interventions can sometimes be anticipated by rational choice theory. However, I also think that it is important to recognize the significant limitations of rational choice in this regard. Most apparently, it is questionable to what extent rational choice theory provides an accurate representation of human decision making (cf. Kahneman, Slovic, and Tversky 1982; Gigerenzer 2000). Experimental results concerning something known as "preference reversal" are particularly relevant to the issue here. The claim that rational choice theory can be used to predict changes to social mechanisms made by policy interventions seems to rest on the following sort of reasoning (cf. Woodward 2000, 220). It is assumed that policy interventions function by changing the information or incentives of agents. Then the claim is that an accurate rational choice model would be invariant under any such intervention. That is because such models presume that any changes to information or incentives affect only agents' beliefs or the consequences associated with particular actions. Meanwhile, agents are assumed to possess stable preferences that are independent of the decision task that reveals these preferences. This assumption is sometimes called "procedure invariance" (Tversky et al. 1990, 204) or "context-free preferences" (Cubitt et al. 2004, 709). Given that one can model changes in incentive structure and the manner in which agents accommodate new information, agents' choices can be derived from their invariant, context-free preference rankings.[4] Thus, the argument concludes, an accurate rational choice model will be invariant under any intervention that targets information or incentives.

The primary weakness of this argument is the assumption of procedure-invariant, context-free preferences. This assumption has been called into doubt by a tradition of experiments on preference reversals (cf. Hausman 1992, chap. 13; Slovic 1995; Guala 2005, 91–108). Classic preference reversal experiments offer subjects a choice between pairs of bets, one that offers a small payoff with high probability and another that offers a large payoff with low probability (cf. Slovic 1995, 365–66). After subjects choose their bet, they are asked to provide a monetary valuation of each. The surprising result is that among those who choose the high probability bet, a substantial proportion gives a higher monetary valuation of the low probability bet. That is, they choose the high probability bet but then say that the low probability bet is worth more to them. Given the natural assumption that people prefer more money to less, such experiments seem to show that a subject's preferences, far from being context-independent, depend on how a choice or task is framed. Some economists have attempted to make preference reversals go away in modified versions of the experiment (cf. Grether and Plott 1979). Others have endeavored to show that the preference reversals are not due to a failure of procedure invariance but can be pinned on some less fundamental assumption implicitly built into the experimental set up (cf. Holt 1986; Segal 1988). However, preference reversals have proved to be an extremely persistent phenomenon, and such attempts to explain them as an experimental artifact have not been successful (cf. Tversky et al. 1990; Camerer 1995, 658–65; Cubitt et al. 2004; Guala 2005, 121–28).

Thus, preference reversal experiments suggest that people often do not have invariant, context-free preferences, but that preferences are frequently constructed on the basis of considerations relevant to particular situations (cf. Slovic 1995; Seidl 2002, 646). If this is correct, then preferences could be affected by changes in information and incentives, since significant changes of these kinds might change the contexts in which preferences are expressed. Thus, interventions affecting information and incentives may change preferences, and without a good understanding of how preferences are constructed in new circumstances, there is no way to predict what the changes will be. However, generalizations about preference reversal phenomena should be made with caution, since questions remain about the relevance of preference reversal experiments to markets and other contexts outside the laboratory walls. Difficulties involved in extrapolating the results of preference reversal experiments will be examined in section 8.3.3.

Even if rational choice theory were a completely accurate representation of human decision making, changes to social mechanisms resulting from policy interventions would still often be difficult to anticipate. That is because changes in social institutions often alter incentives in unexpected ways. Indeed, it is often difficult to accurately characterize the incentives that an existing institution or policy presents to individuals. Thus, even if we can assume that individuals obey rational choice theory, we may nevertheless be ignorant of the mechanisms. The difficulties inherent in attempting to reconstruct a social mechanism from the incentives faced by individuals in a specific social context are amply illustrated by the welfare-reform example discussed in the next section.

Structure-altering interventions, then, are most likely to arise with regard to large-scale, unprecedented policy interventions. Although rational choice theory may sometimes enable one to anticipate changes to a social mechanism that will result from a policy intervention, there are good reasons to suspect that this will often not be possible. Of course, there may be some more adequate theory of human decision making that can address this problem more effectively. But for now, structure-altering interventions are a serious challenge in social science.

8.3 TWO CASE STUDIES

In this section, I examine two cases of extrapolation in social science. The first example is the series of social experiments performed from the 1980s to mid-1990s to assess proposals to change the emphasis of the U.S. welfare system from cash entitlements to temporary assistance and fostering entry into the workforce. This example provides fertile ground for the question of whether mechanisms-based extrapolation is feasible in the social sciences. In fact, the most politically influential methodology did not emphasize mechanisms, focusing instead on randomized experiments that endeavored to evaluate the effects of welfare-to-work programs (cf. Gueron and Pauly 1991; Friedlander and Burtless 1995). Given these experimental results, extrapolation proceeded by simple induction. However, this procedure was challenged by some social scientists who advocated a more traditional econometric approach that endeavored to discover the structure of relationships among the various causes of the outcomes of interest (cf. Manski and Garfinkel 1992). Given the practical impossibility of performing experimental manipulations on any but a few of the variables in such models, this approach invariably relies on inferring causes from observational data.

In section 8.3.1, I propose that the disagreement between the advocates of the structural and experimental approaches is clarified by a more careful distinction between the challenges of (1) estimating a causal effect in a given population and context and (2) extrapolating a causal effect from one population or context to others. The advantages of a randomized controlled experiment are relevant only to (1). Similarly, challenge (2) would still exist even if the causal effect were correctly estimated by nonexperimental means. In short, the dispute is less about the relative merits of experimental and observational methods than about the usefulness of mechanisms for extrapolation in social science. Despite the genuine obstacles to mechanisms-based extrapolation in the present context, inquiries into social mechanisms can provide important qualitative information regarding factors upon which the causal relationship is likely to depend. But such information alone is insufficient for mechanisms-based extrapolation as described in chapters 5 and 6, and is best regarded as providing reasons-to-suppose-otherwise that conscientious simple induction must take into account.

The second case study returns to preference reversal experiments, which were briefly discussed above in section 8.2.2. In this example, there are at least two possible mechanisms that could explain the experimental result, and these mechanisms differ about how pervasive preference reversals are outside the laboratory. Hence, this case study illustrates the simple point that utilizing a mechanisms approach to extrapolation in social science requires greater certainty about mechanisms than often currently exists. The purpose of these two case studies is not to suggest that a mechanisms approach to extrapolation is misguided or unfruitful in social science. Rather, the goal is to clarify the obstacles that must be overcome if mechanisms-based extrapolation is to be useful there.

8.3.1 Extrapolation and Welfare Reform

Imagine that one wishes to evaluate the effectiveness of a welfare-to-work pilot program without the aid of a randomized experiment. One would likely proceed as follows. Suppose that the pilot program is to be implemented at a particular site, wherein a certain number of welfare recipients will participate. One would then search for some suitable comparison group thought to be similar in many relevant respects but not participating in the program. If the two groups were exactly similar except for participation in the pilot program among one and not the other, then the effect the program on, say, earnings could be estimated by the difference in average earnings during the evaluation period. Inevitably, however, there would be differences in the composition of the two groups. The challenge, then, is how to take these into account. The structural approach proceeds by formulating a model that represents the relevant factors influencing earnings and program participation among the groups in question. If this can be done correctly, then the impact of the program can be estimated from statistical data.

Formulating a model of causes of earnings and decisions to participate in a welfare-to-work program requires that thought be given to social processes involved in such things as a person's employment prospects. For example, constructing a model to represent causes of earnings would involve consideration of the labor market in which individuals will be searching for work, what characteristics contribute to success in that market, and so on. As two commentators on this topic put it:

> ... structural evaluation is feasible if one is able to characterize the environment of interest and one understands the social processes at work well enough to permit forecasting with confidence. (Manski and Garfinkel 1992, 11)

There is an obvious similarity between the social processes referred to in this quotation and social mechanisms: both are structured interactions among agents that underlie macrolevel, statistical relationships. Moreover, these authors emphasize the importance of social processes for extrapolation:

> Extrapolation is possible if one is able to characterize the environment of interest and if one understands the social processes that generate program outcomes in this environment. (Ibid.)

The thought here is easy to grasp. The effectiveness of a welfare-to-work program, for instance, depends upon the ability of program participants to find work, and that in turn depends upon such things as the educational level of the participants and the local labor market. And since such factors are likely to vary from one location to another and over time, extrapolating the results of a pilot program by simple induction is problematic. There is, then, a close resemblance between the structural approach to policy evaluation and the mechanisms approach: both insist that attention to processes linking causes to effects is of utmost importance to extrapolation.

But it is questionable whether structural approaches can reliably estimate treatment effects. One objection is that there are typically several models yielding distinct estimates, many of which differ significantly from estimates from randomized experiments (cf. LaLonde 1986; Fraker and Maynard 1987). Models rely on assumptions about such things as the functional form of the relationships between causes and effect, upon the probability distribution of the error terms, and upon whether the error terms are independent. Since the accuracy of such assumptions is often quite uncertain, it is often difficult to assess which model one should choose. A response to such objections is that specification tests can be performed to assess whether the assumptions in the model are accurate (cf. Heckman and Hotz 1989). An example of such a test monitors the matched groups for some time prior to the implementation of the program to see if any significant differences in outcomes emerge. If not, the model passes the specification test. However, some object that such specification tests depend upon questionable assumptions,[5] and are not always effective in weeding out models that generate inaccurate estimates (Friedlander and Robins 1995). Moreover, some studies find that some statistical methods intended to produce balanced comparison groups are often ineffective at reducing bias, and sometimes significantly increase it (Michalopoulos, Bloom, and Hill 2004). Others find that sophisticated matching techniques do far less to reduce bias than simply selecting the comparison group from an adjacent locality (ibid.; Friedlander and Robins 1995).

Consider, then, how an evaluation of a welfare-to-work program would proceed by means of a randomized controlled experiment. From a sample of individuals eligible for the program, some would be randomly selected to participate in it while the remainder would constitute the controls barred from the program. Inevitably, some of those assigned to the program would fail to participate in the activities associated with it. Since restricting attention to the actual participants would undermine the benefits of random assignment, these randomized experiments follow what is known as the *intent-to-treat* methodology. All those assigned to the program are counted among the experimental group whether they participate or not (cf. Friedlander and Burtless 1995, 7). The intent-to-treat methodology makes the variable whose impact is estimated not participation in the program but being officially required to do so. It is arguable that this is not unreasonable, since official requirements are, after all, what policymakers are able to control directly.

A striking feature of this approach is that it does not require that any thought be given to mechanisms relating to employment and earnings of welfare recipients. Likewise, it does not depend upon questionable modeling assumptions, which is a very desirable feature. A related motivation for randomization is that it significantly alleviates the concern that some of the association between cause and effect may be due to the action of an unmeasured common cause (cf. Friedlander and Burtless 1995, 7, 46). These are genuine motivations for using randomized experiments to estimate the impact of pilot welfare-to-work programs. Such considerations were politically influential enough to find their way into a 1988 bill allowing waivers to be granted to states to experiment with welfare-to-work programs: these were to be evaluated by randomized controlled experiments (cf. Manski and Garfinkel 1992, 1). Many of these experiments were carried out by the Manpower Demonstration Research Corporation (MDRC). The MDRC published several studies analyzing the results of these experiments, particularly *From Welfare to Work* (Gueron and Pauly 1991) and *Five Years After: The Long-Term Effects of Welfare to Work Programs* (Friedlander and Burtless 1995). These studies formed a major part of the relevant empirical evidence available to policymakers prior to the 1996 change in the federal welfare program.

The advantages of randomized controlled experiments are relevant to the issue labeled (1) above: the estimation of a causal effect in a population in a particular context. However, the advantages cited in favor of randomization do not pertain to (2), the challenge of extrapolating an estimated causal effect from one population and context to another. The advocates of the experimental approach in the case of welfare reform explicitly note the distinction between estimating a causal effect in a particular context and extrapolating that effect (cf. Gueron and Pauly 1991, 69–70). Nevertheless, they do not attempt to address extrapolation by means of social mechanisms, which is not surprising, given that knowledge of such mechanisms would be difficult to acquire from experiments. Thus, to the extent that their studies license extrapolation, it is only by means of simple induction. Let us consider what motivation there might be for *not* pursuing a mechanisms approach to extrapolation.

The proposal that mechanisms be used as a basis for extrapolation presumes that information concerning mechanisms can be reliably obtained and that the interventions contemplated will not be structure-altering. Both of these premises are doubtful in the present context. My concern with the welfare reform case is chiefly with the possibility that the policy intervention in question would not be structure-altering. In section 8.2.2, I suggested that interventions that are both large-scale and without recent precedents have the greatest potential to be structure-altering. A nationwide, fundamental change in 1996 of welfare programs implemented in the late 1960s is a clear example of such an intervention. Thus, even if reliable knowledge of relevant social mechanisms were present, it is unclear how useful it would be as a basis for extrapolation. In such a context, *not* relying on mechanisms to ground extrapolation might plausibly be regarded as a virtue. Extrapolating the overall effect of the reforms from the experimental outcomes by simple induction could be reliable even if the intervention were structure-altering, just so long as structures were altered in the same manner in the experiment as in the full-scale implementation of the program. If the selection of demonstration sites were also representative of the nation as a whole, then one would have good grounds for an extrapolation by simple induction.

Advocates of the structural approach, however, have argued that the randomized experiment might alter structure differently than the full-scale implementation of the program. James Heckman (1992) suggests that randomization itself might alter the way in which the program is implemented, thereby making the impact in the randomized experiment an unreliable guide to the effect of the program when implemented. For example, this "randomization bias" could arise from the need to expand recruitment of potential program participants in order to achieve sufficiently large experimental and control groups, thereby including ineligible individuals in the program (Heckman 1992, 220–21). A related issue concerns the intent-to-treat methodology employed in the experiments. This procedure undermines extrapolation when rates of non-compliance differ systematically between experiment and implementation of the program. Other authors have noted that important scale effects of the program might not be detectable in experiments evaluating pilot programs. For instance, introducing large numbers of new, unskilled workers into the labor force might make it more difficult for such workers to find employment (Garfinkel, Manski, and Michalopoulos 1992). Yet this "displacement effect" would be unlikely to be detected in a smaller-scale demonstration of the program. The concern about scale effects coincides with observation in section 8.1.2 that large-scale interventions have greater potential to be structure-altering. Given this general rule, it follows that smaller-scale demonstrations of social programs may fail to have the structure-altering characteristics of a full-scale implementation.

There were also some reasons to doubt whether the sample of demonstration sites was indeed representative. Ideally, site selection would have been randomized. However, this was not possible, since sites could not be compelled to perform a randomized experiment to evaluate programs, and many refused to do so for ethical reasons (Hotz 1992, 110–11). The sites that participated in randomized evaluations were those at which officials were willing or could be induced through incentives to do so. Of course, such a sample may fail to be representative. Similarly, the views of those willing to take the time to respond to a survey might not accurately reflect the general state of opinion in a population. In order to assess whether the sample of demonstration sites was representative, one would need to have some understanding of the factors relevant to the effectiveness of the program and their regional distribution.

The concerns outlined in the two foregoing paragraphs raise legitimate questions about extrapolation by simple induction from randomized experiments in the present context. However, the manner in which the issue is framed obscures the important difference between the task of reliably estimating a causal effect in a given context and extrapolating a causal effect. Proponents of the structural approach tend to present extrapolation as the challenge of extending results from experiments to the real world (cf. Manski and Garfinkel 1992, 14–17).[6] Yet barring Heckman's concern about randomization bias and difficulties relating to the intent-to-treat methodology, the challenges for extrapolation described in the foregoing paragraphs arise whether pilot programs are evaluated by observational studies or randomized experiments. The goal of the demonstration evaluations was to provide guidance for policymakers regarding the effects of implementing welfare reform on a national scale. Regardless of how programs at specific sites are assessed, this involves an inference from smaller to larger scales and from specific sites to the nation generally.

In sum, the dispute between the advocates of the structural and experimental approaches to program evaluation is not ultimately about the relative merits of observational studies and randomized experiments for estimating causal effects. Let us suppose, not implausibly, that randomized experiments are indeed superior for this purpose. Even granting this, the central issue that remains is whether the mechanisms-based approach to extrapolation can be usefully employed in social science (and in this context in particular) or whether one is better off relying on simple induction. In the next subsection I consider whether mechanisms-based extrapolation, as described in chapters 5 and 6, would have been feasible in the welfare reform case.

8.3.2 Conscientious Simple Induction

Chapters 5 and 6 explained in detail how mechanisms-based extrapolation can proceed, and illustrated the proposal by means of the aflatoxin example. In this section, I explain how this approach to extrapolation was not likely to have been successful in the welfare reform case. However, this does not entail that mechanisms were irrelevant to extrapolation, since they were necessary for a conscientious and judicious use of simple induction.

A central question about any proposed change to the existing welfare program was whether it would, in general, make those served by the old system better or worse off. Hence, it is plausible to interpret the generalization of interest to extrapolation as a claim about positive causal relevance. The demonstration experiments generally found positive, though rather modest, impacts on income (cf. Friedlander and Burtless 1995). But would these positive impacts extrapolate to a nationally implemented program in potentially less favorable future economic circumstances? The central question in this case, then, is similar to that in the aflatoxin example, wherein the basic question was whether a carcinogenic effect in an animal model could be extrapolated to humans. Recall that the extrapolation in that case was analyzed into two general steps. First, a mechanism through which AFB_1 causes liver cancer was extrapolated from rat to human via comparative process tracing. Next, given that mechanism and some other circumstances plausible in that case—chiefly positive consonance—the extrapolation theorem entailed that the claim concerning positive relevance could be extended to humans. But it is doubtful that either of these steps could proceed similarly in the welfare reform example.

A mechanism found in one population might be significantly different or entirely absent in another. The aim of comparative process tracing, therefore, is to support the inference from the mechanism in the model to the existence of a corresponding (although not necessarily identical) mechanism in the target. This inference proceeded by comparing model and target at stages in the mechanism at which significant differences would likely be present; the greater the similarity at these stages, the firmer the ground for extrapolating the mechanism. Hence, comparative process tracing depends upon reliable background knowledge concerning likely similarities and differences between the model and the target, and a great deal of research in toxicology is consecrated to the acquisition of such information. But carrying out process tracing in the welfare case would run into several obstacles. Not only is it questionable that reliable knowledge concerning likely similarities and differences is available in this case, but it is also unclear how the comparison of stages would proceed. In the aflatoxin example, one could study the metabolism of AFB_1 in rats and compare this with human metabolism of AFB_1 by way of in vitro studies involving cultures of liver cells or blood samples taken from exposed individuals. In the welfare case, one is concerned about extrapolating to larger scales as well as to locales in which some relevant circumstances may differ. The simple problem here is that the operation of a program can be examined only where it is implemented, so that it is unclear that comparative process tracing can facilitate extrapolation to new locations or larger scales.

It is also unlikely that the extrapolation theorem would be of much use in the welfare example. One of the antecedent conditions of that theorem is that the set of mechanisms from cause to effect is positively consonant. Roughly, this means that distinct combinations of mechanisms do not produce conflicting positive and negative effects. Although it is a reasonable assumption in the aflatoxin example, positive consonance is not at all plausible in the case of changes to welfare programs. It is very probable that the new programs would exert a positive impact in some ways and a negative impact in others, and it would be very difficult to specify which of those conflicting impacts would be dominant. Combining these concerns about the applicability of comparative process tracing and the extrapolation theorem with the more general difficulties of structure-altering interventions and reliable learning mechanisms, the prospects of a thoroughgoing mechanisms approach to extrapolation appear rather dim in the case of changes to welfare programs.

However, this does not entail that mechanisms have no relevance to extrapolation whatever in this case. Recall that simple induction was restricted to "related populations" and qualified by a "so long as there is no reason to suppose otherwise" clause. Even if a thoroughgoing mechanisms approach to extrapolation is not possible in the welfare example, examinations of mechanisms may nevertheless help to clarify these aspects of simple induction. For example, it is obvious that the ability of a program to move welfare recipients into stable employment depends crucially upon the local demand for low-skilled labor. Moreover, local labor markets vary significantly from one region to another, as well as over time, in accordance with the business cycle. For example, studies have found very different effects of welfare reform programs in rural and urban contexts (Brown and Lichter 2004). It is also well established that employment among low-skilled workers is more sensitive to the ups and downs of the business cycle than among higher-skilled workers (Hoynes 2000). In addition, there is the possibility of a displacement effect. The displacement effect refers to the consequences for the labor market of introducing large numbers of new, low-skilled job seekers into the workforce. Although unlikely to have much impact on the overall labor market, this infusion of workers may adversely affect wages of less educated women (Bartik 2000, 109). Thus, local supply and demand for low-skilled labor influences the effectiveness of welfare-to-work programs, and the displacement effect points out that the reforms themselves may alter local labor markets to the detriment of those leaving welfare. These considerations place significant limitations on extrapolation by simple induction, and hence must be taken into account in any conscientious application of that inferential strategy.

Neither the influence of local labor markets nor the displacement upon the effectiveness of welfare-to-work programs can be studied by means of randomized controlled experiments. Instead, these topics have been examined by way of observational data analyzed by means of the structural approach described in section 8.3.1. Yet there is reason to think that the primary concerns about social mechanisms as a basis for extrapolation do not undermine the general points described above regarding local labor markets and the displacement effect. Those concerns were that detailed, trustworthy knowledge of mechanisms is often unattainable and that the intervention in question would be structure-altering. However, these concerns are certainly less acute with respect to the claim that the effectiveness of a welfare-to-work program on earnings depends on the local demand for labor. This claim is a simple consequence of the laws of supply and demand, which is about as well established as anything in social science and is unlikely to be altered by a welfare reform program. The proposition that employment among low-skilled labors is more sensitive to the fluctuations of the business cycle, though not a fundamental economic principle, is nevertheless a consistently obtained result (Hoynes 2000, 25, 56–59). Moreover, there seems little reason to suppose that this relationship would be significantly altered by welfare reform. Models predicting the displacement effect do require assumptions regarding some rather uncertain parameters, most importantly concerning the extent to which the infusion of low-skilled female workers would stimulate economic growth and thereby increase demand for such laborers. However, the occurrence of the effect (though not its size) is stable under a range of distinct modeling assumptions. According to Timothy Bartik, the conclusion that the displacement effect is a consequence of welfare reform "can only be avoided if one is willing to assume a labor market that quickly clears and has unusually large labor demand elasticities for less educated women" (2000, 109).

The above discussion illustrates several points about how inquiries concerning social mechanisms can be used to identify important factors upon which a causal relationship depends even when accurate, detailed knowledge of mechanisms is difficult to come by. First, the relevance of some features, such as the local demand for low-skilled labor and sensitivity to the business cycle, is readily derivable from firmly established economic relationships that are unlikely to be altered by welfare reform. The analysis of the displacement effect does depend upon assumptions about difficult-to-estimate elasticities in the demand for low-skilled female labor. But in this case, the displacement effect is robust under a wide range of likely values of this parameter. Second, the implications of the analyses for welfare reform are qualitative. Welfare-to-work programs have a smaller positive impact on earnings in depressed labor markets; large-scale reductions in welfare rolls will exert a downward pressure on wages for less educated women. Any *quantitative* estimate of the size of these effects would inevitably require debatable assumptions of the sort described in section 8.3.1. Nevertheless, the qualitative information about how the effects of the welfare reform program are likely to vary according to local labor market conditions and fluctuations in the business cycle has clear implications for responsible policymaking.[7]

8.3.3 Preference Reversals in the Real World

Experiments concerning preference reversals are one of the examples used by Guala to illustrate methodological problems relating to extrapolation. Guala writes:

> Economists nowadays generally agree that PRs [preference reversals] are a real laboratory phenomenon rather than a mere illusion of the instruments of observation. The second phase of research began when experimenters turned their attention to the robustness of reversals *outside* the laboratory. (2005, 225)

In this section, I examine this extrapolation problem from the perspective of the mechanisms approach proposed here. Unlike the welfare reform example, questions about the extrapolation of results of preference reversal experiments are not tied to an unprecedented, large-scale policy intervention. Thus structure-altering interventions are a less acute problem in the preference reversal than in the welfare reform case. The primary challenge confronting the extrapolation of the results of preference reversal experiments is uncertainty about the explanation of those results. Some plausible mechanisms suggest that preference reversals are widespread in real life contexts, while others suggest that preference reversals are far more limited. I illustrate this point by reference to two possible explanations of preference reversals.

Recall that "preference reversal" refers to the following sort of situation. A subject is asked to consider a pair of bets: one that has a high probability of a small payoff (the P-bet) and another that has a low probability of a high payoff (the $-bet). The subject is asked to choose one of the pair, and then to provide a monetary valuation of each. A preference reversal is said to occur if the bet not chosen is given a higher monetary valuation than the one that was chosen. The "expected" or "predicted" preference reversal occurs when the subject chooses the P-bet but gives a higher monetary valuation to the $-bet. It is important to notice that, without further elaboration, such an outcome is not necessarily anomalous or even particularly surprising. For example, it is hardly news that a merchant might overprice her goods in the hopes of garnering additional profits. Hence, if the subject were actually selling a bet on the market, it might be rational for her to strategically overprice it.

A number of procedures have been designed to avoid such strategic pricing effects. The most commonly utilized method in preference reversal experiments is what is known as the Becker-DeGroot-Marschak (BDM) elicitation method. The BDM elicitation method works in the following manner (cf. Roth 1995, 19–20). The subject is given a lottery (e.g., the $-bet) and asked the price at which she would be willing to sell it, on the following conditions:

(1) The asking price a will be compared with a randomly generated buying price b;

(2) If $b \geq a$, then the subject exchanges the lottery for b;
(3) If $a > b$, then the subject keeps and immediately plays the lottery.

The rationale for the BDM elicitation method is characterized by Alvin Roth as follows:

> It is not hard to see that the dominant strategy for a utility maximizer faced with such a mechanism is to state his true selling price (i.e., the price that makes him indifferent between keeping the lottery and selling it). (1995, 20)

To see the reasoning here, let p represent the subject's true selling price. Suppose that the subject chooses an asking price a strictly greater than p. But then if $a > b > p$, the subject forgoes the opportunity for an advantageous exchange. Suppose then that the subject sets $a < p$. Then if $p > b > a$, the subject must sell the lottery for less than its value to her. Thus, the subject should set $a = p$, that is, her asking price should reveal her true selling price. Notice that it is important that the buying price b is randomly generated. This means that the asking price has no effect on how much the buyer will offer to pay, something that is usually not true in real markets. Moreover, the subject is not allowed to search for further buyers if the initial buying price is not to her liking. In contrast, a real-life merchant may continue to search for other buyers if initial offers do not meet the price she demands.

Much of the initial response to preference reversal experiments consisted of ingenious proposals about how the BDM elicitation method itself might be responsible for the effect. However, since the results of preference reversal experiments were subsequently replicated with a variety of elicitation methods, such explanations of the preference reversal experiment are no longer regarded as very promising (Seidl 2002, 634). According to a different explanation proposed by Amos Tversky, Paul Slovic, and Daniel Kahneman, "The primary cause of PR [preference reversal] is the failure of procedure invariance, especially the overpricing of low-probability, high-payoff bets" (1990, 204). Procedure invariance "requires strategically equivalent methods of elicitation to yield the same preference order" (ibid.). Thus, in the context of the experiment, the choice between the P-bet and the $-bet is strategically equivalent to providing a monetary valuation (or monetary ranking) of the bets. In such a context, to choose one bet while giving a higher monetary valuation to the other is a violation of procedure invariance. Both of the explanations that I consider associate preference reversals with failures of procedure invariance, but they differ about how widely the phenomenon should be expected outside the laboratory walls.

The first explanation I will discuss is the one proposed by Tversky, Slovic, and Kahneman, namely, that preference reversals are due to scale compatibility (1990, 211). In general, scale compatibility is the hypothesis that "the weight of any aspect (for example, probability, payoff) of an object of evaluation is enhanced by compatibility with the response (for example, choice, pricing)" (ibid.). Thus, when choosing between the P-bet and the $-bet, subjects focus on the probability, and accordingly take the P-bet. But when asked to provide a monetary valuation, subjects focus on the monetary scale, and hence rank the $-bet more highly. If this explanation is correct, then preference reversals certainly cannot be dismissed as some oddity of the laboratory with no relevance to real life. Preferences are expressed in various ways, of which overt choices and prices are merely two common examples. Thus, if preference orderings are indeed so closely linked to the scale on which the options must be ranked, then preference reversals should be a prevalent feature of everyday life.

But scale compatibility is not the only possible explanation of preference reversals. Consider a person in a preference reversal experiment. If the person is risk-averse, then she will choose the P-bet when given the choice of playing one or the other. But now consider this question: Which bet has a higher monetary value? The subject might naturally interpret this question as a request for an estimate of the market value of each bet. Indeed, outside the laboratory walls, questions about prices normally are requests for information about market rather than personal value. For concreteness, suppose that the P-bet gives a 99 percent chance of winning $10, while the $-bet gives a 0.099 percent chance of winning $10,000. While it is obvious that no one will pay more than $10 for the P-bet, it is likely that some risk-seeking individuals will agree to pay more than $10 for the $-bet. Hence, in a real market, a seller can expect to obtain a higher price for the $-bet than for the P-bet. Consequently, it is understandable that the subject might give a higher monetary valuation for the $-bet, even if she would play the P-bet when given a choice between the two.

The possibility that this simple insight might explain preference reversals is developed in a paper by Xiaoyong Chai (2005). Let us use the expression "market price reversal" to refer to the hypothesis that subjects in preference reversal experiments often interpret questions about the prices of bets as questions about their market value. Market price reversal can explain a puzzling aspect of the data found from some of the earliest experiments. It turns out that expected preference reversals are far more common when subjects are asked to specify a *selling* price rather than a *buying* price (cf. Seidl 2002, 622–23), a result that Chai replicates (2005, 190). A question about the selling price is very likely to be interpreted as a question about market value, whereas a question about buying price is more likely to suggest an assessment of personal value to the subject. Moreover, as Chai observes, a prospective buyer of the $-bet might hope to find a risk-averse person who possesses it and who is willing to sell it for less than its expected value (2005, 182). Consequently, market preference reversal predicts that "unexpected" preference reversals (in which the subject chooses the $-bet but prices the P-bet higher) should be more common when a buying rather than a selling price is asked for, which is indeed correct (Seidl 2002, 623; Chai 2005, 191). In contrast, the scale compatibility hypothesis does not explain why the frequency of expected and unexpected preference reversals is linked to whether a selling or a buying price is requested. For posing a question about price in terms of selling rather than buying does not change the scale on which the valuation is made.

The primary objection to the market price reversal hypothesis is that elicitation procedures like BDM are designed precisely for the purpose of eliminating market-based considerations. A utility-maximizing subject will, in response to the BDM procedure, give her own true price of the lottery, rather than an estimate of its market price. And indeed, the rate of expected preference reversals is significantly less in the BDM elicitation method than when subjects are merely asked to specify a selling price (cf. Seidl 2002, 623). However, there are several reasons why BDM and other such elicitation methods might not eliminate market pricing effects. The first is simply that many subjects may not appreciate the strategic implications of the BDM procedure. The subject is required to quickly learn the new rules of the game played in the experiment, and may not have time to carefully devise an optimal strategy in response to them. Under such circumstances, a subject might follow a heuristic that works reasonably well in a real-world context that appears similar to the experimental situation. Thus, many subjects may still propose selling prices that are estimates of market values. Moreover, as Chai observes, in the BDM procedure "the absence of a human buyer is *unverifiable* to the subjects" (2005, 185). That means that the independence between posted selling price and offered buying price is not something that the subject can directly verify herself. In light of this, Chai modifies the BDM mechanism so that the buying price offered is present as physical money in a sealed envelope placed before the subject (2005, 186–87). The subject, therefore, can easily see that what selling price she posts has no effect on what buying price is offered. Chai found that the difference in relative frequency between expected and unexpected preference reversals was not statistically significant when the envelope method was used (2005, 191).

Market price reversal, therefore, attributes preference reversals to a systematic difference between real markets and experimental circumstances. In real markets it is normally the case that (a) sellers are free to search out buyers (and vice versa) and (b) the posted selling price is not independent of the buying price offered. Under these circumstances, choosing between options is not strategically equivalent to specifying prices for them. In contrast, neither (a) nor (b) obtains in the experimental context, and choosing and pricing are strategically equivalent. According to the market price reversal hypothesis, then, preference reversals in experiments can in fact be regarded as violation of procedure invariance, since subjects do give conflicting rankings to strategically equivalent options. But that result does not entail that violations of procedure invariance are widespread in economic or other social contexts. For if market price reversal is the correct explanation, these experiments show only that it is possible to trick people with decision scenarios that differ strategically from the situations in ordinary life that they superficially resemble.

At present it remains an open question whether scale compatibility or market price reversal is the better explanation of the results of preference reversal experiments. The pertinent point for the purposes of this book is that the extent to which these results can be extrapolated outside of the laboratory depends crucially on which explanation is correct. Of course, that conclusion should hardly be surprising from the perspective of a mechanisms approach to extrapolation. If extrapolation relies on information about mechanisms, then what can be extrapolated may depend on what the relevant mechanisms are.

8.4 CONCLUSION

In this chapter, I have inquired into the prospects of utilizing mechanisms-based extrapolation in social science by reference to a pair of extended case studies. These case studies illustrated two central challenges to this extension of methodology, which have to do with the difficulty of obtaining reliable information concerning social mechanisms and the potential that interventions will be structure-altering. The potential for interventions to be structure-altering is prominent in the welfare reform example, wherein one wished to extrapolate the impact of a large-scale policy reform on the basis of pilot experiments. I argued that mechanism-based extrapolation along the lines of the aflatoxin example was unlikely to have been feasible in this instance. Nevertheless, inquiries into social mechanisms are helpful in the welfare reform case insofar as they provide information needed for a conscientious application of simple induction. The case study concerning preference reversals highlighted a second challenge for mechanisms-based extrapolation in social science, namely, uncertainty about the operative mechanisms. In that example, one plausible mechanism suggests that preference reversals are widespread outside the laboratory, while another suggests that they are more limited. In the next chapter, I further explore the discovery of social mechanisms and its relation to causal inference more generally.

9

Social Mechanisms and Process Tracing

The examples discussed in the previous chapter illustrated that uncertainty about mechanisms is a central challenge for a mechanisms approach to extrapolation in social science. The discovery of mechanisms is part of the broader issue of causal inference, but at the same time several authors have maintained that examining social mechanisms can significantly ameliorate the challenges confronting causal inference in social science (Elster 1983, 47–48; Little 1991, 24–25; Hedström and Swedberg 1998, 9). The discovery of social mechanisms, therefore, can be properly addressed only in the context of causal inference in social science more generally. Claims about the importance of mechanisms for causal inference rest on the observation that in social science, it is often impossible to discern the correct causal hypotheses on the basis of statistical data alone. Thus, process tracing, a method specifically devised for the discovery of mechanisms, is sometimes advanced as an additional means for narrowing the set of possible causal explanations (cf. George and Bennet 2005, 214–15, 223). But it is far from clear that process tracing differs from causal inference from statistical data in any significant way, and if so, just how. For a mechanism is more than a collection of contiguous objects: a mechanism essentially involves a pattern of causal interactions. And how is one to distinguish causal interaction from mere physical contiguity except by reference to statistical regularities? In this chapter, I develop an account of process tracing that addresses these concerns.

I begin by explaining the challenge for causal inference from observational data. Next, I critically examine accounts concerning the manner in which mechanisms allegedly assist causal inference in social science. It is sometimes asserted that reliable causal inference in social science is *impossible* without knowledge of mechanisms, a proposition that Kincaid (1996, chap. 5) has disputed. Although I agree with Kincaid in rejecting the claim that mechanisms are always required for causal inference in social science, I also maintain that the proposal can be made independently of that proposition. On the interpretation I suggest, the account of how mechanisms assist causal inference in social science has a positive and a negative aspect. On the positive side, we can infer that X is a cause of Y if we know that there is a mechanism through which X influences Y. The negative flip side is that if no plausible mechanism running from X to Y can be conceived of, then it is safe to conclude that X does not cause Y. As I explain, neither of these two theses entails that mechanisms are necessary for causal inference in social science, and consequently they are not undermined by the criticisms raised by Kincaid. Nevertheless, I argue that this account of how mechanisms assist causal inference in social science is not successful as it stands. The positive account is not helpful unless some explanation is given of how it is possible to learn about social mechanisms despite the challenges for causal inference from observational data. Yet no such explanation has hitherto been provided. On the other hand, the effectiveness of the negative side is undermined by the ease of imagining mechanisms connecting nearly any two variables representing aspects of social phenomena.

I provide an analysis of process tracing that aims to shore up the positive side of the argument. It is sometimes claimed that process tracing is fundamentally distinct from causal inference from statistical data (cf. George and Bennett 2005, 207). However, the appropriate contrast with process tracing is not inference from statistical data, but rather what I call *direct causal inference*. Suppose one wishes to learn the causal relationships among a set of variables that represent macrofeatures of a complex system, for instance, the impact of federal deficits on interest rates and economic growth. One strategy is to collect a large sample of statistical data concerning these variables; given these data, one attempts to draw conclusions about the causal relationships among them. In contrast, process tracing attempts to use knowledge of causal generalizations about the system's components together with information concerning their configuration to infer mechanisms relating macrolevel variables. Process tracing, then, presumes that the macrobehavior of the system can be reconstructed from interactions of its parts and that causal knowledge about the components may be more directly accessible than about macrofeatures of the system. On this account, process tracing is not separate from inference from statistical data, since statistical data are needed to discover causal generalizations concerning the mechanism components. Nevertheless, process tracing may be able to produce results when direct causal inference alone is unable to yield informative conclusions.

9.1 CONFOUNDERS AND INSTRUMENTAL VARIABLES

A great deal of social science involves collecting statistical data relevant to some phenomena of interest (e.g., through government records or surveys) and performing tests to decide whether pairs of variables are probabilistically dependent conditional on sets of other variables. In some cases, the purpose of such inquiries might be solely to identify factors that can serve as useful forecasting tools, but often the goal is to discover what variables cause which others. In the social sciences, this leads to the thorny problem of making causal inferences without the aid of experiment. The obstacle to such inferences is that there are often several possible causal hypotheses capable of explaining the statistical data. In particular, without experiment a probabilistic dependence between two variables might be explained either by one variable being a cause of the other or by the existence of a common cause of both. We can call this the *problem of confounders*, where the term "confounders" refers to common causes, often unmeasured, that might explain an observed correlation.

There are several proposals concerning how one can reliably learn causal relationships from statistical data even when unmeasured common causes may be present. Such proposals identify a specific set of favorable conditions in which causal structure may be reliably inferred from statistical data, the problem of confounders notwithstanding. The question concerning such proposals is twofold. First, how commonly do the favorable circumstances specified by the method occur? Second, how could one know whether those circumstances obtained in a given case? The first question asks how useful the method can be in general, while the second asks how the method's suitability in a particular instance can be assessed. A comprehensive survey of methods of causal inference from observational studies is obviously far beyond the scope of this chapter. I consider one example that illustrates the issues, namely, the method of instrumental variables.

One important issue in discussions of causal inference concerns premises linking causation and things that serve as evidence for it. If the evidence consists of statistical data, then these premises primarily concern the relationship between probability and causality. That is, one estimates probabilities from statistical data and then draws inferences about causal relationships from the probabilities. One of the most important principles relating probabilities and causation is the causal Markov condition (CMC). The CMC asserts that, conditional on its direct causes, a variable is probabilistically independent of any set of other variables that does not include any of its effects. We encountered the CMC in section 4.4.1 when discussing the disruption principle. As noted there, an important consequence of the CMC is the "screening-off" rule. In the graph on the left in Figure 9.1, X and Y are related only as effects of the common cause Z. Thus, the CMC entails that X and Y are probabilistically independent, conditional on Z in the graph on the left, and likewise in the graph on the right. The CMC also entails the principle of the common cause, which asserts that if X and Y are probabilistically dependent, then X is a cause of Y, Y a cause of X, or there is a common cause of the two. For example, the CMC entails that X and Y are probabilistically independent in the graph in Figure 9.2.

However, the CMC does *not* entail that X and Y are independent conditional on Z in Figure 9.2. In general, conditioning on a collider

Figure 9.1 An illustration of the causal Markov condition

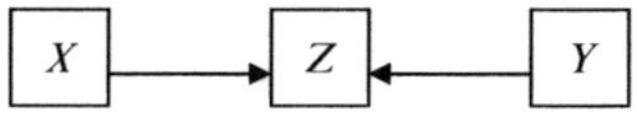

Figure 9.2 A collider

(that is, a node with two arrows pointing into it) or an effect of a collider may induce probabilistic dependence where it would otherwise be absent. To see intuitively why this should be so, suppose that *X* indicates the quantity of gasoline in your car's tank; *Y*, whether the battery is charged; and *Z*, whether your car starts. Without knowing anything about *Z*, *X* and *Y* are independent. However, suppose that your car does not start. In this situation, the information that there is gas in the tank suggests that the battery is likely the culprit.[1] There is a graphical concept called d-separation that enables one to read off the independence relationships entailed by the CMC from any directed acyclic graph. (D-separation is explained in the Appendix, and may be helpful for understanding some of the ensuing discussion.)

The motivation for the CMC is that it is true of any acyclic causal system with independent error or disturbances (Steel 2005). The second of these conditions—independent error terms—is of the greatest concern for our purposes. Typically, it is not satisfied for sets of variables that one actually measures in an observational study, owing to the existence of unmeasured common causes. Thus, the CMC would generally be invoked in the following way: the causal structure relating the measured variables can be embedded in a more extensive structure that satisfies the CMC. This assumption entails, for instance, that any probabilistic dependence among measured variables not arising from causal connections among themselves indicates the presence of an unmeasured common cause.

A second important principle relating probability and causality is the faithfulness condition (FC), which asserts that the *only* probabilistic independence relationships in acyclic causal structures are those entailed by the CMC. For instance, the FC entails that *X* and *Y* are probabilistically dependent in each graph in Figure 9.1, and that *X* and *Y* are probabilistically dependent, conditional on *Z* in the graph in Figure 9.2. The grounds for the FC were discussed in section 4.4.2. The CMC and FC are rarely explicitly stated in scientific research, but they are pervasively assumed. For example, in observational studies it is standard practice to statistically control (e.g., by linear regression) for possible common causes of a pair of variables of interest. The residual correlation is then tentatively attributed to the direct influence of the suspected cause. Clearly, such a procedure assumes the screening-off rule illustrated in Figure 9.1. Similarly, the FC is implicit in nearly any study claiming that there is "no link" between a certain pair of pair variables. That is, such studies typically report that no statistically significant association was found among a pair of variables, and thus conclude that neither is a cause of the other.

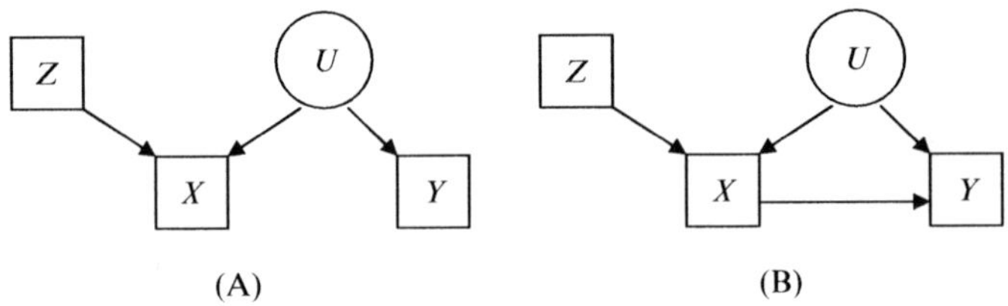

Figure 9.3 An instrument variable

The CMC and FC are also implicit in the method of instrumental variables. Suppose that one wishes to assess the effect of X upon Y, but suspects that unmeasured common causes are present. In that case, the effect of X upon Y cannot be directly inferred from the probabilistic dependence between them, since some or even all of that dependence might be due to the confounders. However, suppose it is known that the variable Z is a cause of X but otherwise unrelated to Y, as in the two graphs in Figure 9.3. The variable U represents unmeasured common causes of X and Y (squares indicate measured variables; circles, unmeasured ones). Both graphs predict a probabilistic dependence between X and Y, but the presence of Z makes it possible to distinguish between these two alternatives by means of statistical data. If graph (A) is correct, then it follows from the CMC that Z and Y are independent. In contrast, if graph (B) is right, then the FC entails that Z and Y are probabilistically dependent.

In the above example, Z is an instrumental variable with respect to X and Y. The concept of an instrumental variable can be defined as follows. Let us say that Z is *exogenous with respect* to X and Y just in case Z is neither an effect of nor shares a common cause with either of these two variables. Then Z is an instrumental variable with respect to X and Y exactly if (1) Z is a cause of X, (2) Z is exogenous with respect to X and Y, and (3) any directed path from Z to Y passes through X. Condition (3) is sometimes called the *exclusion restriction* (Angrist, Imbens, and Rubin 1996, 447; Rosenbaum 2002, 181). In the two graphs in Figure 9.4, Z fails conditions (2) and (3), respectively. In graph (A), there is a common cause of Z and Y, and hence Z is not exogenous. In graph (B), there is a directed path from Z to Y that does not pass through X (namely, $Z \rightarrow U \rightarrow Y$), so Z fails the

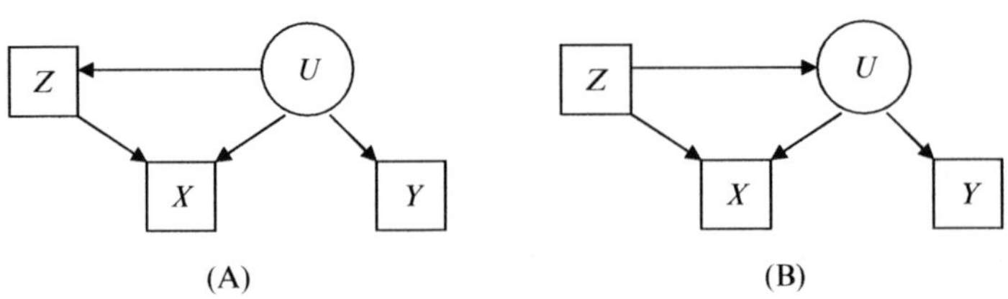

Figure 9.4 Not an instrument variable

exclusion restriction. Given the FC, both graphs predict that Z and Y are probabilistically dependent, despite the fact that X is a cause of Y in neither. Consequently, a probabilistic dependence between Z and Y does not show that X is a cause of Y if either condition (2) or (3) fails.

The presence of an instrumental variable, therefore, is an example of a favorable circumstance that facilitates causal inference from statistical data. Indeed, under appropriate conditions, the covariance between Z and Y divided by the covariance between Z and X is a consistent estimator of the impact of X upon Y.[2] But the difficulty lies in establishing a bona fide instrumental variable. Uncertainty about the exclusion restriction is the most significant problem in this regard.

One common application of the method of instrumental variables occurs in randomized experiments with non-compliance. Recall that in the experiments designed to evaluate welfare-to-work programs, not all those assigned to participate in the program actually did so. Nevertheless, assignment might be an instrumental variable with respect to program participation and income. Assignment to the program is clearly a cause of participation in it, and assignment is exogenous thanks to randomization. However, the exclusion restriction is more problematic. For example, recipients might interpret assignment to the program as an indication that their benefits will soon be terminated, and this perception might stimulate them to search more actively for employment.

In a double-blind randomized experiment, the exclusion restriction is on much firmer ground. For instance, a placebo ensures that any psychological impact resulting directly from treatment assignment is distributed equally among control and experimental groups. But when double-blinds are absent, as is the case in welfare-to-work experiments, the exclusion restriction is often uncertain. Concerns about the exclusion restriction are also prominent in examples of putative instrumental variables outside of randomized experiments. One well-known alleged instrumental variable is draft number with respect to military service in Vietnam and subsequent income. Since Vietnam era draft numbers were assigned randomly, they satisfy items (1) and (2) of the definition of an instrumental variable, but the exclusion restriction is again uncertain (cf. Angrist, Imbens, and Rubin 1996, 452). For instance, since draft deferments could be obtained by attending university, it is plausible that draft number might influence the choice to attend college, and hence earnings (cf. Angrist 1990, 330).

Is there, then, any statistical test for whether a putative instrumental variable Z satisfies the exclusion restriction with regard to a pair of variables X and Y? One approach is to provide data against particular hypotheses about how the alleged instrumental variable might fail the exclusion restriction (cf. Angrist 1990, 330; Angrist and Krueger 1992, 334–35). The problem with this strategy is that it is difficult to know whether all of the ways the exclusion restriction could fail have been considered. A different suggestion is that if Z is a genuine instrumental

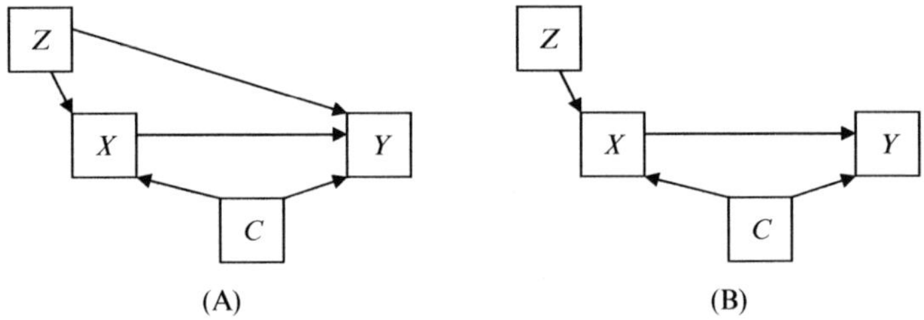

Figure 9.5 Testing the exclusion restriction

variable, then it should be independent of Y, conditional on X and measured causes of Y (cf. Heckman 1996, 460). To see the idea, consider the two graphs in Figure 9.5. In graph (B), Z is an instrumental variable with respect to X and Y, but not in graph (A), owing to the arrow from Z to Y, which violates the exclusion restriction. However, given the CMC and the FC, these two graphs make differing predictions about conditional probabilities. Since X is a cause of Y, both graphs predict that Z and Y are probabilistically dependent. Moreover, both predict that Z and Y are dependent conditional on X. This is because X is a collider on the path $Z \rightarrow X \leftarrow C \rightarrow Y$, and hence conditioning on it induces probabilistic dependence, as noted with regard to Figure 9.2. However, conditional on *both X and C*, graph (A) predicts that Z and Y are probabilistically *dependent*, while graph (B) predicts the opposite. Thus, it seems that the exclusion restriction should be accepted if Z and Y are independent, conditional on X and C, and rejected otherwise.

Unfortunately, there is a serious problem with this test. In particular, a genuine instrumental variable can be expected to fail the test when there is an unmeasured common cause of X and Y. To see why, consider the DAG in Figure 9.6. Owing to the path $Z \rightarrow X \leftarrow U \rightarrow Y$, this graph predicts that Z and Y are probabilistically dependent conditional on X and C (recall that conditioning on colliders induces probabilistic dependence, as explained with regard to Figure 9.2). Nevertheless, Z is an instrumental variable with regard to X and Y. Thus, the problem of confounders is a significant obstacle for causal inference from statistical data without experiment. When unmeasured common causes are present, association and conditional association are unreliable indicators of causal influence.

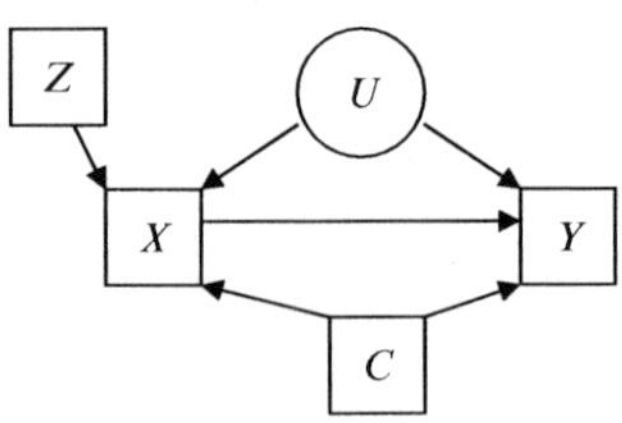

Figure 9.6 An instrument variable that fails the test

Moreover, the presence of unmeasured common causes makes it very difficult to decide whether a putative instrumental variable satisfies the exclusion restriction.

9.2 MECHANISMS TO THE RESCUE?

Advocates of social mechanisms are motivated in large measure by concern about the problem of confounders, a difficulty sometimes referred to as "spurious correlation" (cf. Elster 1983, 47). Mechanisms are sometimes advanced as the basis of a partial solution of this problem. For example, according to Daniel Little:

> We can best exclude the possibility of a spurious correlation between variables by forming a hypothesis about the mechanisms at work in the circumstances. If we conclude that there is no plausible mechanism linking nicotine stains to lung cancer, then we can also conclude that the observed correlation is spurious. (1991, 24–25)

The argument in this passage rests upon the following principle:

> (M) *X* is a cause of *Y* if and only if there is a mechanism from *X* to *Y*.[3]

Clearly, (M) is not intended as a universally true principle regarding causality, since there is presumably some "rock bottom" level of physical causation below which no mechanisms lie. Thus, (M) should be understood as being restricted to complex systems composed of multiple, interacting components, for instance, an organism or a society. In effect, (M) amounts to the claim that mechanisms are equivalent to causal structure. Although a very natural assumption, this identification is by no means self-evident, as discussions in chapters 3 and 8 show.

Given (M), if there is a mechanism from *X* to *Y*, then *X* is a cause of *Y*. Conversely, if there is no mechanism from *X* to *Y*, then *X* is not a cause of *Y*, regardless of any statistical association between them. The latter of these two corollaries of (M) is illustrated by Little's example about nicotine stains and lung cancer. Peter Hedström and Richard Swedberg make the same point with a different example:

> Some epidemiological studies have found an empirical association between exposure to electromagnetic fields and childhood leukemia. However, the weight of these empirical results is severely reduced by the fact that there exists no known biological mechanism that can explain how low-frequency magnetic fields could possibly induce cancer.... The lack of a plausible mechanism increases the likelihood that the weak and rather unsystematic empirical evidence reported in this epidemiological literature simply reflects unmeasured confounding factors rather than a genuine cause relationship. (1998, 9)

Jon Elster (1983, 47–48) provides a similar example, though with a slight twist that we will consider below.

9.2.1 Kincaid's Objections

Although the proposal just outlined maintains that inquiries into social mechanisms can significantly ameliorate the problem of confounders, it did not assert that they are the *only* means for resolving this problem. However, that claim is sometimes made in close association with the argument described above. For example, according to Little, "It is *only* on the basis of hypotheses about underlying causal mechanisms that social scientists will be able to use empirical evidence to establish causal connections" (1995a, 53–54; italics added). Kincaid raises two objections to the claim that causal inference is possible in social science only when a mechanism has been identified (1996, 179–82). The first of these objections takes the form of a reductio ad absurdum.

In Kincaid's discussion, the proposition at issue is that "we need to identify individualist mechanisms to confirm causal relations between social variables" (1996, 179).[4] Let us formulate this proposition in the following way:

> (M*) One knows that X is a cause of Y only if at least one mechanism from X to Y can be identified.

Kincaid's reductio ad absurdum then proceeds as follows. Suppose that a mechanism relating two macrolevel social variables is demanded to support the claim that one of the variables is a cause of the other.

> Do we need it at the small-group level or the individual level? If the latter, why stop there? We can, for example, always ask what mechanism brings about individual behavior. So we are off to find neurological mechanisms, then biochemical, and so on. (Ibid.)

Given (M*), therefore, demands for mechanisms can be pressed all the way down to fundamental physics, yielding the absurd result that no causal claim can be established unless such impossible amounts of detail are provided.[5]

One might try to defend (M*) from such objections by maintaining that it is intended to apply only to fields in which controlled experiment is not possible and not all common causes can be measured (Little 1998, 10–12). Since controlled experiments are routine in such fields as psychology, neuroscience, and molecular biology, Kincaid's reductio ad absurdum would be blocked. Although there is some merit to this response to Kincaid's reductio ad absurdum, (M*) is nevertheless quite problematic. For example, it is *sometimes* possible to perform good randomized, controlled experiments in social science, and it is *sometimes* the case that one has a bona fide instrumental variable (cf. Angrist and Krueger 1991, 1992). Thus, it is better to simply agree with Kincaid that (M*) is false, but to point out that it is not required for the proposal described in the foregoing section. That proposal rested on the proposition (M), which stated that X is a cause of Y if and only if there is a mechanism from X to Y. The target of

Kincaid's reductio ad absurdum, meanwhile, is (M*). But does (M) entail (M*)? Defenders of mechanisms in social science sometimes seem to presume that it does. Consider the following statement by Little:

> I maintain that the central idea of causal ascription is the idea of a causal mechanism: to assert that A causes B is to assert that A in the context of typical causal fields brings about B through a specific mechanism (or increases the probability of the occurrence of B). This may be called "causal realism," since it rests on the assumption that there are real causal powers underlying causal relations. This approach places central focus on the idea of a causal mechanism: *to identify a causal relation between two kinds of events or conditions, we need to identify the typical causal mechanisms through which the first kind brings about the second.* (1995, 34; italics added)

Notice that the first sentence in this quotation is a statement of (M): there is a causal relationship just in case there is an underlying mechanism. In contrast, the italicized sentence is a statement of (M*): mechanisms must be identified before we can claim to know that one variable is a cause of another. However, (M) does *not* entail (M*).

To see the point, imagine a person who accepts (M) but also regards randomized controlled experiments as a reliable means of learning about cause and effect. Suppose that a randomized controlled experiment establishes that X is a cause of Y. Then the person concludes from (M) that there is a mechanism from X to Y. Nevertheless, the person may not be able to identify any mechanism from X to Y; in short, she knows *that* there is a mechanism, but not *what* this mechanism is. Therefore, such a person would not be committed to the proposition that is the basis of Kincaid's reductio ad absurdum, that is, she would not be committed to (M*). Her inability to identify a mechanism is compatible with her knowledge that there is a mechanism and, hence, with her knowledge of a causal relationship. In general, one can consistently accept (M) while rejecting (M*) by holding, reasonably enough, that tracing mechanisms is not the only possible way to learn about cause and effect.

Kincaid's second objection to (M*) is that there are ways of distinguishing between cause and mere correlation available to social scientists that have nothing to do with mechanisms, particularly by conditioning on potential confounders (1996, 179–80). As the discussion of instrumental variables illustrated, there are indeed favorable circumstances in which causal conclusions can be inferred from statistical data without experiment and perhaps without knowledge of mechanisms linking cause and effect. However, that does not undermine the proposal that mechanisms significantly aid causal inference in the social sciences, since the favorable circumstances may occur rarely and be difficult to recognize when present. Furthermore, the suggestion that one statistically control for all common causes is not very helpful, given that the inability to exhaustively consider all potential common causes is a basic element of the problem of confounders.

Indeed, evidence regarding conditional dependencies and independencies may fail to unambiguously identify causal structure even when all potential common causes have been measured. For example, consider the two graphs in Figure 9.1. Both of these graphs predict that *X* and *Y* are probabilistically dependent. Moreover, they both predict that *X* and *Y* are independent conditional on *Z*. So measuring *Z* would not enable us to decide whether *X* is a cause of *Y*, even if there were no other confounders.[6] Elster's argument for the importance of mechanisms to causal inference in social science is motivated by an example that illustrates the same point (1983, 48). In Elster's example, the variable *X* represents "the percentage of female employees who are married" and *Y* represents "the average number of absences per week per employee" (ibid.). Elster supposes that *X* and *Y* are positively correlated but that they are independent conditional on a third variable *Z*, "the amount of housework performed per week per employee" (ibid.). Both causal graphs in Figure 9.1 can explain this imagined statistical evidence; hence, we are unable to decide from that evidence alone whether *X* is a cause of *Y*. However, Elster suggests, since there is no plausible mechanism through which *Z* could influence *X*, we can conclude that *Z* is not a cause of *X*, and hence not a common cause of *X* and *Y*. The only remaining alternative, therefore, is that *X* is a cause of *Z*, which in turn is a cause of *Y*. Thus, this example illustrates how (M) might be used to establish a positive causal conclusion that could not have been reached through the examination of statistical data alone.

In sum, although Kincaid is correct that (M*) is false, that proposition is not required for the account presented above of how inquiries into mechanisms play a central role in causal inference in the social sciences. Nevertheless, that proposal leaves much to be desired.

9.2.2 The Positive and Negative Sides

There is, as was noted above, a positive side and a negative side of the account of the importance of mechanisms to causal inference in the social sciences. The positive side rests on the premise that we can show that *X* is a cause of *Y* if we can discover a mechanism from *X* to *Y*. The negative side relies on the premise that we can infer that *X* is not a cause of *Y* if we know that there is no mechanism from *X* to *Y*. It was the negative side that was illustrated by Little's nicotine-stains-and-lung-cancer example, Hedström and Swedberg's example concerning electromagnetic fields and childhood leukemia, and Elster's example about the proportion of female employees and number of missed workdays. But the negative side of the account of the importance of mechanisms to causal inference in social science is very problematic.

The problem lies in the ease of imagining social mechanisms through which nearly any macrolevel social variable can influence another. It is rarely the case that no plausible mechanism can be imagined that could connect two variables representing aspects of social phenomena.

Consider, for instance, a well-known example from the sociological literature discussed by one of the contributors to Hedström and Swedberg's (1998) volume, Diego Gambetta. The example is the negative correlation between satisfaction and opportunity for advancement among military personnel, reported in Samuel Stouffer's *The American Soldier* (1949). Surprisingly, soldiers in branches of the military offering little opportunity, such as the military police, were on average more satisfied with their positions than those in branches with greater chances for advancement, such as the Army Air Corps. Gambetta describes five mechanisms proposed by sociologists over the years to account for how greater opportunity could cause less satisfaction (1998, 114–19). However, he does not consider the alternative possibility that opportunity has little or no negative influence on happiness, and that the association found by Stouffer is due to an unmeasured common cause. For example, it is possible that ambitious people are much more likely to embark on career paths that promise greater opportunities for advancement and that their lofty aspirations are also more likely to make them dissatisfied with their current station in life. Listing possible mechanisms through which opportunity could produce unhappiness does nothing to rule out this plausible alternative. Indeed, this case illustrates how an overabundance of plausible mechanisms is a major source of difficulty for causal inference in the social sciences.

No doubt there are some pairs of variables X and Y representing collective aspects of social phenomena such that no plausible mechanism through which X causes Y can be imagined. However, I suspect that such cases are too few and far between for the no-plausible-mechanism strategy to be of much use in distinguishing cause from mere correlation in social science. Although Elster, Little, and Hedström and Swedberg each illustrate their argument with an example, only Elster's—a toy example not based on actual research—has any relation to social science. Despite their interest in doing so, these authors apparently found it difficult to produce a serious example of actual social research in which the inability to imagine a plausible mechanism from one social variable to another significantly aided causal inference.[7]

As we saw, (M) can be used to generate a positive as well as a negative account of the value of mechanisms to causal inference in social science. Having found the negative proposal wanting, let us turn to the positive one. From (M) it follows that if we know that there is a mechanism from X to Y, we can infer that X is a cause of Y. The difficulty is that it is unclear how we are to learn about mechanisms in a way that does not run directly into the problem of confounders, which was the problem that mechanisms were supposed to help us overcome. For example, consider Little's discussion of how one acquires knowledge of mechanisms:

> To credibly identify causal mechanisms we must employ one of two forms of inference. First, we may use a deductive approach, establishing causal

> connections between social factors based on a theory of the underlying process. ...Second, we may use a broadly inductive approach, justifying the claim that **a** caused **b** on the ground that events of type **A** are commonly associated with events of type **B**.... But in either case the strength of the causal assertion depends on the discovery of a regular association between event types. (1991, 30)

Thus, according to Little, the identification of causal mechanisms depends on prior knowledge of probabilistic dependencies among variables.[8] But the problem of confounders immediately rears its ugly head at this juncture, since the probabilistic dependence might result from a common cause rather than from A being a cause of B.

The same difficulty confronts an account of process tracing given by Alexander George and Andrew Bennett (2005, chapter 10). George and Bennett use an analogy about a row of dominoes to illustrate their account of process tracing (2005, 206–7). Imagine that you are shown a series of dominoes lined up in a row. You then leave the room, and when you return, the first and last dominoes are lying flat and the intermediate ones are concealed behind a screen. In order to know whether the toppling of the first domino caused the last to fall, it is necessary to lift the screen to see if the intermediate dominoes are also toppled in the appropriate direction. Lifting the screen to check the positions of the intermediate dominoes is the analogue to process tracing as understood by George and Bennett. According to this proposal, process tracing is a method for testing hypotheses about the causes of a particular event, what would be called "token" or "actual" causes in the philosophical literature (cf. Eells 1991; Pearl 2000; Halpern and Pearl 2005). The domino example draws attention to the fact that hypotheses about the actual causes of a particular outcome often have implications for what events occurred between the (alleged) cause and effect. In George and Bennett's account, then, process tracing is a method of testing hypotheses about actual causes by investigating whether the predicted sequence of intermediate events indeed occurred.

George and Bennett's proposal is very sensible, but leaves precisely same issue unresolved that Little's did. That is, the presence of the sequence of predicted events between the alleged cause and the effect is not sufficient to establish actual causation—it is also necessary to show that the sequence *is not a coincidence*. In other words, the chain of intermediate events must be causal: each event in the chain is the actual cause of the subsequent one. But to show that the sequence of events is not merely coincidental, as Little observes, one needs to appeal to some causal generalization. And it is difficult to see how causal generalizations could be learned without some type of inference from statistical data. Thus, George and Bennett's account of process tracing does not indicate how mechanisms can be discovered without already having resolved the challenges confronting causal inference from statistical data in social science.

The positive side of the proposal, then, stands in need of some explanation of why the problem of confounders is less acute when it comes to learning mechanisms than it is for macrocausal relationships in the system. That is the task that I undertake in what follows.

9.3 PROCESS TRACING

In Chapter 5, process tracing was characterized as instantiating a mechanism schema by means of tracing forward or backward, where the components and interactions at one stage place restrictions on those at preceding and subsequent stages. For example, if one discovers that a particular type of cancerous tumor results from a specific mutation, then an earlier stage of the mechanism must involve something capable of producing a mutation of just that sort. At this level of description, however, it is not clear how process tracing differs from other methods of inferring causal relationships from statistical data, possibly in conjunction with background knowledge. Consequently, it is not clear how process tracing ameliorates the problem of confounders. In this section, I provide an account of process tracing that aims to address these concerns.

9.3.1 Direct Versus Indirect Causal Inference

In order to properly understand process tracing, it is important to be clear about its intended contrast. It is sometimes said that process tracing is utterly distinct from methods that endeavor to draw causal inferences from statistical data. For example, George and Bennett write, "Process-tracing is fundamentally different from methods based on covariance or comparisons across cases" (2005, 207). In the foregoing section, I argued to the contrary that process tracing is inextricably intertwined with causal inference from statistical data. The appropriate distinction, I suggest, is not between one method that relies on statistical data and another that can proceed independently of such information. Rather, the distinction is between what I call direct and indirect causal inference. Direct causal inference attempts to infer the causal relationships among a set of variables by examining the probabilistic relations among *those same* variables. By contrast, indirect causal inference attempts to learn the causal relationships among a set of variables by examining the causal relations among a *distinct yet related* set. In process tracing, the distinct yet related variables represent features of component parts of the larger system of interest. The usefulness of process tracing, then, rests on the possibility that the causal relationships among the components are more directly accessible than those among the macrofeatures of the system. Let us consider this idea in more detail.

Suppose that one is interested in the causal relationships among a set of variables **V** that represent macrofeatures of a system S. The system might be an economy, an organism, or a machine. The variables in **V** might represent such things as inflation and unemployment if S is an

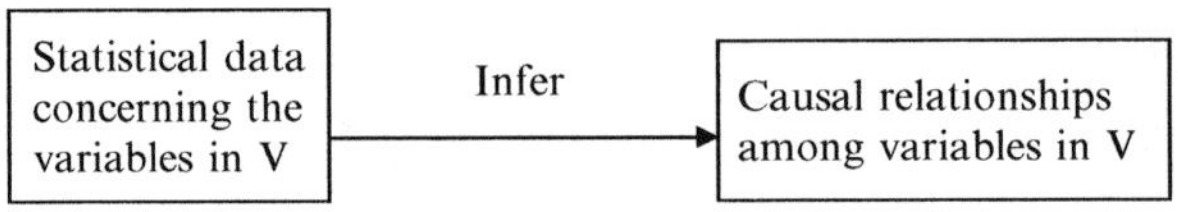

Figure 9.7 Direct causal interference

economy, or exposure to aflatoxin B_1 and liver cancer if S is a person. One strategy for learning about the causal relationships among the variables in **V** is by means of statistical data concerning those variables. I call this *direct causal inference* (or *direct inference* for short), since the strategy focuses directly on the variables of interest and the probabilistic relations among them. Direct inference can be represented schematically as shown in Figure 9.7.

For example, suppose that **V** contains variables representing federal deficits, inflation, economic growth, interest rates, and unemployment. Suppose, moreover, that the chief concern is to estimate the effect of federal deficits on economic growth. Then direct causal inference might proceed by comparing carefully matched periods that differ with respect to federal deficits. Attempting to infer the causal relationships among the variables in **V** from statistical data concerning them together with the CMC and FC would also fall into the category of direct causal inference. The method of instrumental variables is direct inference with one wrinkle: an instrumental variable is sought, and if a promising candidate is found, it is added to **V**.

Process tracing does not focus directly upon the statistical relationships among the variables in **V**, but rather upon the components of S and their configuration. This can be depicted schematically as in Figure 9.8.

Of course, direct inference and process tracing are not mutually exclusive: both could contribute to knowledge of the causal relationships among the variables in **V**.[9] Moreover, direct inference will almost certainly be an important source of knowledge of causal generalizations concerning the components. However, that inference would involve a set of variables distinct from **V**. Let **C** be a set of variables representing features of the components. Process tracing, then, exploits the possibility that the causal relationships among **C** may be more easily learned than those among **V**. One way this could be is if it is possible to perform experiments on the components, but not the system as a whole. For example, experimental economists can perform randomized experiments

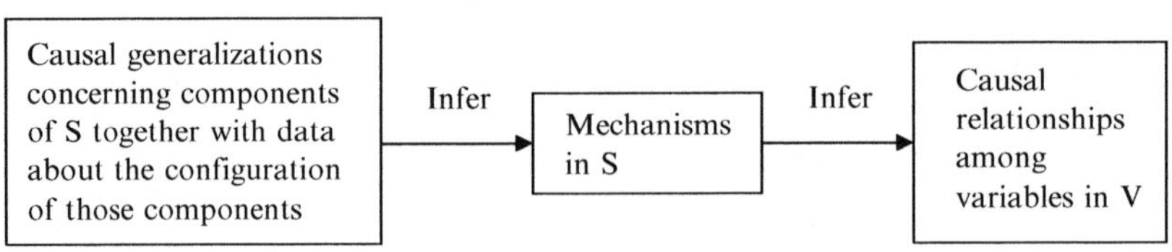

Figure 9.8 Process tracing

involving individuals or small groups but not entire economies. Similarly, ethical considerations prohibit an experiment in which persons are exposed to aflatoxin B_1, yet it is possible to experimentally study, say, the metabolism of that compound in vitro by means of cell cultures. Even when experiments cannot be performed on the component parts of the system, there may be better observational data with regard to the relevant features of the components than for the system as a whole. Or it may be that the possible confounders have been more exhaustively listed and measured with regard to the components than for the macrofeatures of the system. In short, there may be a variety of practical reasons why the causal relationships among the variables in **C** can be more directly ascertained than among those in **V**.

Let us examine a case of process tracing in social science. For example, consider Malinowski's hypothesis that the possession of many wives was a cause of wealth and influence among Trobriand chiefs (1935). Malinowski's evidence for this hypothesis is primarily nonstatistical; it consists of descriptions of social processes in Trobriand society. First, there is a custom whereby brothers contribute substantial gifts of yams to the households of their married sisters—gifts that are larger than usual when the sister is married to a chief. Second, political endeavors and public projects undertaken by chiefs are financed primarily with yams. As this case illustrates, process tracing in social science often provides evidence for the existence of several prevalent social practices that, when linked together, constitute a mechanism. Supposing that Malinowski was right about the two features of Trobriand society just described, the conclusion that the number of wives had an influence upon wealth among Trobriand chiefs is unavoidable.

Let us consider how this example fits into the abstract outline of process tracing depicted in Figure 9.8. The system in this case would be Trobriand society of the early twentieth century, and the set **V** would include variables indicating social status, wealth, and number of wives. The components would be the individual Trobrianders, categorized as brothers-in-law, wives, and chiefs. Given these components, process tracing utilizes causal generalizations concerning them and information about their configuration to infer a mechanism. The causal generalizations in this case would mostly be psychological, for instance, concerning human aspirations for wealth and social status. The configuration of the components would include the salient relationships among the relevant groups (e.g., brothers-in-law are required to give yams to sister's household) as well as the preferences and beliefs typical of members of these groups (e.g., Trobriand men wish to be regarded as good farmers and generous in giving yams). Such a configuration constitutes what one might call a practice or custom. In the Trobriand case, for example, it was a custom for brothers-in-law to provide a sizable contribution of yams to the households of their married sisters. Recall that a social mechanism consists of agents grouped into categories associated with characteristic modes of behavior in such a way as to generate a macrolevel regularity.[10] Identifying a set of practices that link together to form a social mechanism, then, constitutes a successful application of process tracing.

Process tracing is most noticeable in cases in which good statistical data are not available. Consider Brian Ferguson's account of the effect of the introduction of such manufactured items as steel tools upon indigenous warfare in the context of colonial expansion and in the political consolidation of postcolonial states, particularly among the Yanomami.[11] Ferguson has long argued that European colonial expansion profoundly reshaped indigenous warfare in the Americas and elsewhere (cf. Ferguson 1990). One of the ways in which European contact influenced warfare was through introduction of manufactured valuables, particularly such steel tools as machetes, axes, and knives. These items were often quick to become necessities of life, but they differed significantly from their indigenous analogues in that they could not be manufactured locally. Moreover, in more than a few cases, these precious items were available only from a limited number of peripheral source points. Ferguson argues (1984, 1995) that in such circumstances, groups close to the source often attempted to establish a local monopoly on the flow of manufactured goods so as to trade on advantageous terms with their neighbors. Naturally, such monopolizing efforts often generated resentment among more remote groups, which might attempt to circumvent the monopolists or dislodge them by force. Likewise, the would-be monopolists might resort to violence to maintain their privileged position.

Ferguson's proposals concerning the effect of manufactured valuables on indigenous warfare, although controversial, have been taken seriously,[12] and the major themes of his arguments have been taken up by other authors (cf. Reedy-Maschner and Maschner 1999; Steel 1999). The data in such ethnohistorical studies are typically of a very fragmentary nature: reports of missionaries, explorers, ethnographers, and recollections of elderly informants. Even when such information is reliable, it rarely suffices for anything resembling a sophisticated statistical analysis. Not surprisingly, then, process tracing plays an important role in the causal arguments in such circumstances.

Of course, process tracing is not limited to situations in which no reliable statistical data are available. Consider John Donohue and Steven Levitt's (2001) essay, "The Impact of Legalized Abortion on Crime." Donohue and Levitt argue that the legalization of abortion in the United States following the 1973 *Roe v. Wade* decision is the most significant factor responsible for the decline in U.S. crime rates in the 1990s. Although it may seem surprising that legalizing abortion could affect crime rates two decades later, Donohue and Levitt suggest a plausible mechanism linking the two (2001, 386–89). Women choose to have an abortion when the child would be unwanted, for example, because they would be unable to adequately care for and economically support it. Donohue and Levitt cite a variety of studies that report correlations between being raised in adverse family situations and criminality in early adulthood (2001, 388–89). Thus, they propose that the legalization of abortion in 1973 resulted in a birth cohort that, when entering its prime crime age eighteen to twent-four years later, contained a smaller proportion of individuals disposed to criminal behavior. Donohue and Levitt give several lines of statistical evidence for this hypothesis. For example, they show that the drop in crime rates occurred earlier in states that legalized abortion prior to *Roe v. Wade*, and that the initial decrease occurred in categories of crime disproportionately committed by those in the eighteen-twenty-four age group (2001, 395–99). Not only does Donohue and Levitt's study illustrate the combination of process tracing and causal inference based on statistical data, it also illustrates the role of statistical data in process tracing itself. For example, the causal generalization that unwanted children are more likely to become criminals is obviously a proposition that must be tested by reference to statistical data. Moreover, the relevant evidence with regard to this generalization is not limited to data relating to the relationship concerning the legalization of abortion in the United States in 1973 and the subsequent drop in crime rates there in the 1990s. For example, Donohue and Levitt cite several studies from eastern Europe and Scandinavia which found that children born to women who were denied access to abortions were more likely to engage in criminal behavior (2001, 388). This is an example of how a more extensive set of data may be available with regard to the behavior of components of a system than with regard to the macrolevel features of the system. As explained above, that is one of the chief motivations for process tracing as a research strategy.

I regard process tracing as both a procedure for developing, or formulating, causal hypotheses and for providing evidence for them.[13] Process tracing is not intended merely as a means of inventing intriguing new hypotheses, and it is clear that it must be more than this if it is to ameliorate the problem of confounders. For if the *only* evidence for hypotheses generated through process tracing consisted of statistical tests concerning macrolevel variables, the problem of confounders would be confronted anew with no progress having been made toward its resolution. After all, the difficulty lies not in imagining hypotheses concerning the causes of social phenomena, but in deciding which among the large number of such conceivable hypotheses is correct. Therefore, it is important to address concerns that a skeptic might have concerning the ability of process tracing to provide compelling evidence for causal claims.

9.3.2 Objections Considered

A striking feature of Malinowski's account of the relationship between number of wives and chiefly power in Trobriand society is that it is compelling, yet utterly lacking in statistical sophistication of any kind. No large sample of data is produced to demonstrate a positive correlation between wealth and number of wives among Trobriand chiefs. Nor is any thought given to alternative hypotheses that could generate such a probabilistic dependence if it existed. For example, it might be that wealth is a cause of having many wives, and not vice versa. Or perhaps establishing alliances with other chiefs results both in having more wives (used as a means of cementing political bonds) and in greater wealth. As if by magic, Malinowski seems to have established that one variable is the cause of another without the aid of any experimental or statistical technique for dealing with the possibility of unmeasured common causes. All this might make a skeptic wonder whether process tracing is too good to be true. Surely, the skeptic might object, the problem of alternative hypotheses capable of accounting for the available evidence is not made to disappear through a description of social institutions. Moreover, the skeptic could continue, the obstacles to reliably learning social institutions and the implications of their joint operation through process tracing appear at least as formidable as the challenges confronting direct causal inference. Let us consider these objections.

To begin with, it is important to emphasize how modest the accomplishments claimed for process tracing often are. Without the aid of statistical data, the best one can hope to establish by means of process tracing are purely qualitative causal claims. For instance, in the Malinowski example, all we can conclude is that there is at least one path through which the number of wives exerts a positive influence upon wealth among Trobriand chiefs. Not only does this conclusion fail to specify anything about the strength of the influence generated by this mechanism, it does not even entail that the overall effect of the number of wives upon wealth is positive. One would naturally presume that having more wives would mean having more members of the household to provide for, which would be expected to exert a downward influence upon wealth. Statistical data concerning the average cost-benefit ratio in yams of acquiring additional wives would be needed to decide which of these two conflicting influences was predominant, and no such data is provided by Malinowski. Thus, a successful application of process tracing allows one to conclude that a mechanism is present from one variable to another, but this information alone tells one very little about probabilistic causal relationships. This situation is very similar to a case in which one has successfully extrapolated a mechanism from a model organism by means of comparative process tracing (see section 5.3). Given *only* that there is a mechanism from cause to effect in the target population, one knows very little about the probabilistic impact of the cause—for instance, whether it raises or lowers the probability of effect overall.

Yet that process tracing, on its own, is only intended to establish qualitative causal conclusions may not fully allay the skeptic's suspicions. Section 9.1 illustrated just how difficult causal inference from statistical data can be. Can such formidable challenges really be overcome by a relatively unsophisticated method like process tracing? The answer to this concern, I suggest, lies in the fact that it may be possible to learn causal relationships concerning the components of a system even when the causal relationships among the macrofeatures of that system cannot be discovered by direct inference. Consider the difficulties confronting direct causal inference in the Malinowski example. Suppose that $\mathbf{V} = \{W, S, N\}$, where W, S, and N are variables indicating wealth, social status, and number of wives, respectively. It is quite plausible that there is a causal connection between each pair of these variables that is unmediated by the third. For example, status is likely causally linked to wealth as cause or effect independently of number of wives. Likewise, it is likely that status and number of wives are linked by a path that is not mediated by wealth. Finally, it is likely that greater wealth and number of wives would be linked by a path unmediated by status. If all this were indeed the case, then the FC would entail that there are no (nontrivial) probabilistic independencies among these variables. That is, W and S would likely be probabilistically dependent both marginally and conditional on N, and likewise for the other two combinations of variables. The point here may be clarified by reference to causal graphs representing plausible alternatives in this case. Given the FC, each of the graphs in Figure 9.9 predicts that there are no marginal or conditional independencies among the measured variables. For example, all predict that W and N are probabilistically dependent conditional on S, that S and W are dependent conditional on N, and so on. Yet the graphs disagree about the causal relationship between number of wives and wealth. So, if one of these graphs were the correct one, no amount of statistical data concerning the variables W, S, and N could tell us whether number of wives is a cause of wealth.

What this example illustrates is that ability of direct inference to yield informative conclusions depends upon which causal structure is actually present. Some structures generate patterns of independence and conditional independence not generated by alternatives that differ with regard to the question at issue. Such structures have, as it were, a probabilistic fingerprint that reveals useful information about causal relationships. Other structures generate patterns of independence and conditional independence that are also generated by a class of alternatives. In such cases, direct inference is unable to produce informative causal conclusions. The Malinowski example seems likely to be an instance of this latter, thornier sort.

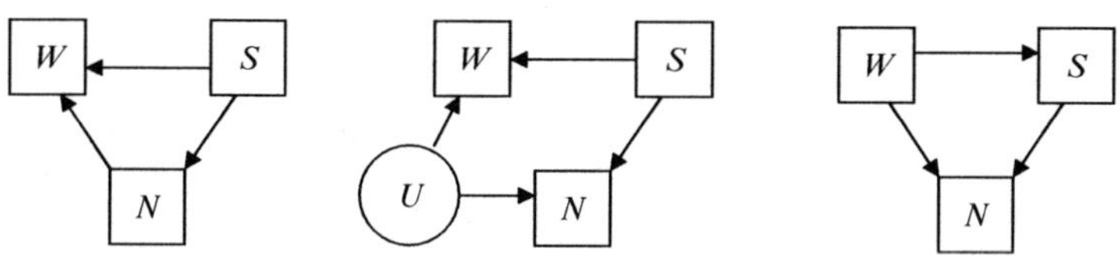

Figure 9.9 Statistically indistinguishable alternatives

One might try to remedy this situation by means of an instrumental variable. For instance, the British colonial authorities banned polygamy for moral reasons, so one might attempt to assess the effect of *N* upon *W* through comparing the wealth of the chiefs before and after the ban went into effect. However, it is highly questionable whether the ban would qualify as an instrumental variable. Although the ban was certainly a cause of *N* and possibly exogenous with respect to *N* and *W*, it is very doubtful that the exclusion restriction is satisfied, since the ban was accompanied by a variety of actions aimed at undermining the chiefs, who after all were rivals to the British colonial authorities. In sum, it is doubtful that Malinowski or anyone else would have been able to draw informative conclusions by direct inference alone, even if a large sample of good statistical data had been available.

Underdetermination also confronts process tracing, but with regard to a distinct set of variables whose causal relations can often be studied by distinct means. In the case of Malinowski's hypothesis concerning marriage, yams, and chiefly power, the central difficulty is that of interpreting a social practice. Malinowski faced the challenge of making an inference about a social practice of which he had no initial inkling from beginning observations of large quantities of yams being moved to and fro. No doubt, multiple possible explanations occurred to Malinowski at this point. Evidence relevant to distinguishing between these alternatives would typically consist of observing people's behavior and asking them about what they are doing and why, and what would happen to someone who behaved differently. Thus, Malinowski makes observations about the quantity of yams produced by several apparently typical men, and he makes observations about the quantity that is contributed to the households of sisters. In addition, he questions native informants about the process, relying in part on what-would-happen-if questions such as "What would people say if so-and-so did not contribute a significant portion of his crop to the households of his sisters?" Since the two underdetermination problems are distinct, it is possible that there are situations in which one of them is resolved while the other is not. Hence, Malinowski might have successfully used process tracing to establish the existence of a social mechanism through which the number of wives influenced wealth among Trobriand chiefs while having no solution to the challenges confronting direct inference.

Of course, inferences concerning which interpretation is best depend on substantive causal generalizations about human psychological and cognitive tendencies. The usefulness of process tracing, therefore, depends upon knowledge of such generalizations being more directly accessible than those concerning variables in **V**. Some simple psychological generalizations can be plausibly regarded as obvious background knowledge, and I suspect that such generalizations often suffice for relatively straightforward interpretations of social practices as in the Malinowski example. But I agree with Todd Jones (1999, 356–58) that there are interpretations in which less obvious psychological generalizations would be called for. In such cases, Jones's proposal is that one should turn to modern cognitive psychology for assistance. Although generalizations from such a source cannot be regarded as obvious background knowledge, controlled experiments are much more frequently a practical possibility in cognitive psychology than in social science. This illustrates the point made in the foregoing section that process tracing relies upon causal relationships among the components being more directly accessible than those at the macrolevel.

Nevertheless, the skeptic is certainly correct that process tracing can be hampered by uncertainty concerning interpretation. There is no denying that trustworthy understandings of social practices are sometimes hard to come by. In broad outlines, the situation of process tracing that relies on interpretation of social practices resembles that for direct inference insofar as it is it is most effective under certain favorable circumstances. With respect to interpreting social practices, these favorable circumstances would include the following:

(1) The practice in question is exhibited in publicly accessible settings.
(2) There is no prohibition, taboo, or other obstacle to open discussion of the practice.
(3) The practice is transparent to participants, in the sense that participants have a reasonably clear understanding of its functioning.

Conditions (2) and (3) facilitate learning what participants regard as the rules and practices, while (1) allows for comparisons with actual behavior. All of these three conditions appear to be satisfied in the Malinowski example, but circumstances in other cases are not as favorable. Philosophical worries about interpretation in social science often focus on attributions of symbolic meaning that would not occur to the participants, and hence in which (3) is not satisfied (cf. Martin 1993; Jones 1998, 1999). In Ferguson's account of Yanomami warfare (described in the preceding section), failures of (1) and (2) pose real difficulties. Those planning violent acts often deliberate in private and attempt to carry out assassinations in as clandestine a manner as possible. Moreover, they typically insist upon socially acceptable, self-serving justifications of their actions, for example, that an assassination was retribution for some past wrong inflicted by the victim. Of course, those committing acts of violence may come to believe their own rationalizations, in which case failures of (2) shade into failures of (3).

Whether the favorable circumstances that facilitate process tracing in social science are more or less widespread than their counterparts in the case of direct inference is difficult to know. But the important point for our purposes is that these two sets of favorable circumstances are potentially independent. Favorable conditions for process tracing may be present while those for direct causal inference are absent, and vice versa. In some especially fortuitous cases, favorable circumstances for both may co-occur, while in other unlucky situations both may be lacking. Process tracing, then, extends the class of cases in which informative causal conclusions can be drawn in social science into not uncommon situations in which direct inference alone would bear little fruit. That is what I think is right about the intuition that mechanisms are of central importance to causal inference in social science. However, that conclusion does not support the claim that mechanisms are a sine qua non for causal inference in social science. For there are some cases in which favorable circumstances allowing for direct causal inference are present in social science. Moreover, direct inference may be applicable in some cases in which process tracing is not particularly helpful. Thus, the correct image is not of one method that is more fundamental than the other, but rather of two mutually supporting approaches.

9.4 CONCLUSION

This chapter has two closely interrelated aims: to explore how social mechanisms can be discovered, and how such inquiries can ameliorate challenges confronting causal inference from statistical data in social science. I proposed that accounts hitherto provided for the usefulness of mechanisms for causal inference in social science can be interpreted so as to be independent of the proposition that causal inference is *never* possible without mechanisms, a proposition rightly critiqued by Kincaid. Nevertheless, I argued that even given this more charitable interpretation, the proposal still faces serious challenges. The negative side of the argument is undermined by the ease of imagining plausible mechanisms that could link nearly any two macrolevel social variables. The positive side of the argument is ineffective unless some explanation is provided of how knowledge of mechanisms can be acquired in a way that avoids the challenges facing causal inference from statistical data in social science, particularly the problem confounders. Yet advocates of social mechanisms have not provided any such explanation.

Consequently, I developed an account concerning how the positive side of the argument could be improved, based on the notion of process tracing. Process tracing was distinguished from what I termed direct causal inference, wherein one endeavors to discover causal relationships among a set of variables by examining the statistical relationships among them. In contrast, process tracing endeavors to infer mechanisms underlying those statistical relationships from the configuration of components of the system and causal generalizations concerning those components. Thus, process tracing exploits the possibility that causal knowledge concerning the components of a system may be more directly accessible than of its macrofeatures. Both process tracing and direct inference are useful when certain favorable circumstances are present. But since the favorable circumstances of the two approaches are potentially independent, process tracing may enable informative causal conclusions to be drawn in cases in which direct inference alone would not. Claims about the importance of mechanisms for causal inference in social science, therefore, are best understood as maintaining that this type of situation is quite common.

10

Looking Back and Ahead

In this brief concluding chapter, I encapsulate some take-home messages of this work and sketch some open questions. I begin with the most general themes, and then turn to main conclusions regarding more specific topics.

Extrapolation in heterogeneous populations is a pervasive challenge in the biological and social sciences that is linked to policy questions about the regulation of toxic substances and the reform of social programs. Although extrapolation is of obvious methodological interest in toxicology and experimental economics, I hope to have demonstrated that it is also deeply intertwined with philosophical issues associated with biology and social science, including causation, ceteris paribus laws, and reductionism. There are several ways of approaching extrapolation, of which simple induction, capacities or causal powers, and mechanisms were considered in detail. Simple induction is certainly a part of the story, but it is also limited in some crucial respects, as was illustrated by reference to cases of animal extrapolation in toxicology. Hence, it is desirable to find some means of extrapolation capable of surmounting obstacles confronting simple induction. Any such proposal must confront the extrapolator's circle and must explain how extrapolation can be justified even when there are some causally relevant differences between the model and the target. I argued that hitherto proposed versions of the capacities and mechanisms proposals do not adequately address either of these challenges. Consequently, I tried to do better by means of a further development of the mechanisms approach.

Chapters 3, 4, 5, and 6 were dedicated to exploring and clarifying the premises that underlie the mechanisms approach and how they are capable of supporting extrapolation in particular cases. The take-home messages here are more specific than the broad themes just adumbrated. Some underlying premises of mechanisms-based extrapolation are what might be called *basic presuppositions*: conditions without which the enterprise would stand little chance of success. Among these I would include the identification of mechanisms with causal structure and the disruption principle. According to the first of these, mechanisms are that which generate probability distributions and indicate how those distributions change given interventions. This identification, however, does not provide any details about the nature of the link between mechanisms and probability, nor about how and when interventions change probabilities. The disruption principle provides an important link of this kind. It says that interventions on X change the probability distribution of Y just in case there is an undisrupted mechanism from X to Y. The disruption principle, therefore, is an important premise in an account of how mechanisms can ground the extrapolation of probabilistic causal claims.

In addition to these basic presuppositions, there what can be called *facilitating conditions*: circumstances that, when present, facilitate extrapolation. Several circumstances of this kind were discussed. The most important, I think, is knowledge of likely sources of similarity and difference between mechanisms in model and target. This information is needed for comparative process tracing, which aims to use a mechanism in a model as a basis for inferring the existence of a corresponding mechanism in a target. Given that such an inference is possible, one typically still desires to know whether the overall impact of the cause in the target is to raise or lower the chance of the effect. Consonance asserts that distinct combinations of mechanisms present in the target population do not generate conflicting positive and negative influences. When available, this condition greatly facilitates extrapolation of claims about causal relevance, for instance, that exposure to a compound increases the chance of liver cancer. In Chapters 5 and 6, I examined a biological example—the case of aflatoxin B_1—in which both the basic presuppositions and the facilitating conditions were very plausible. Chapter 8 considered prospects of utilizing mechanisms-based extrapolation in social science. The mechanism designed for the FCC broadcasting license auctions with the aid of experimental economics (briefly discussed in section 8.1) is probably one of the best social science cases of comparative process tracing. However, the two other case studies examined in Chapter 8 illustrated obstacles that can confront mechanisms-based extrapolation in that domain. In the welfare reform example, there was a very serious possibility that the intervention of interest would be structure-altering with regard to the mechanisms. In the preference reversal case, uncertainty about the mechanism made extrapolation similarly uncertain, since one mechanism suggests that the phenomenon is widespread outside the laboratory while the other suggested that it is far less so.

The preference reversal example led directly to the discussion, in Chapter 9, of *process tracing*, which has been proposed by several authors as a means for discovering mechanisms. It is sometimes claimed that without mechanisms it is rarely (if ever) possible in social science to distinguish cause from mere correlation. I argued that preexisting accounts of how inquiries into mechanisms aid causal inference are inadequate as they stand. I maintained that the appropriate contrast with process tracing is not causal inference from statistical data, but rather what I termed *direct causal inference*. Suppose one wishes to learn the causal relationships among a set of variables, say, education, socioeconomic status of parents, and income. Direct inference endeavors to estimate those relationships from statistical relationships among the variables in question. In contrast, process tracing examines the components of the system whose features those variables represent. Process tracing is motivated by the possibility that causal generalizations concerning the components may be more easily learned than those representing macrolevel properties. Given this, process tracing attempts to reconstruct macrolevel relationships from the configuration and interactions of the components. Both direct inference and process tracing can yield informative conclusions under certain favorable circumstances. But since the favorable circumstances in each case are potentially independent, process tracing may enable informative causal conclusions to be drawn in cases where direct inference alone could not (and vice versa).

An account of extrapolation should provide insight into related methodological and philosophical issues. One obviously relevant methodological dispute concerns whether animal models can serve as a basis for extrapolation to humans, or merely as sources of interesting hypotheses to be tested by clinical and epidemiological studies. I showed how my account of extrapolation provides a diagnosis of flaws in arguments claiming to show that animal models cannot support informative causal conclusions about humans. A consequence of my position is that, to the extent that the ethics of animal research depends upon methodological questions, across-the-board ethical arguments vindicating or condemning it are not likely to be cogent. Rather, such arguments must pay careful attention to case-specific details. Extrapolation is also intimately linked to the question of ceteris paribus laws, generalizations qualified by a clause to the effect of "other things being equal" or "so long as nothing interferes." I showed how difficulties confronting the most problematic interpretation of "ceteris paribus" stem from the assumption that this phrase qualifies a universal generalization. These difficulties vanish if "ceteris paribus" is understood in reference to an inference schema specifying sufficient conditions for extrapolating a claim about causal relevance. The mechanisms approach to extrapolation is also linked to reductionism insofar as it is committed to what I termed *corrective asymmetry*. I also explained how corrective asymmetry can be used as a criterion for a form of reductionism that is consistent with pluralism.

Finally, I would like to close by sketching some open questions that are suggested by the discussions contained in the foregoing chapters of this book. Although many such questions may have occurred to the reader, I mention only these.

- The aflatoxin B_1 example nicely fit the account of mechanisms-based extrapolation expounded here. How representative is this example of other biological cases? To what extent do the challenges confronting mechanisms-based extrapolation described in Chapter 8 also arise in biology?
- What is the potential usefulness of the mechanisms approach to extrapolation in social science? Would the approach work, provided there was more adequate knowledge of social mechanisms,

perhaps combined with advances in cognitive and social psychology? Will one ultimately have to conclude that this is a methodological approach that, though valuable in biology, is ill suited to social science?

- Comparative process tracing was presented in Chapter 5 in an entirely informal manner. Is there some way to integrate comparative process tracing within a more general and precisely articulated approach to causal inference, such as Bayesian networks? What new insights about extrapolation would ensue from this?

As these questions indicate, this book has hardly answered all of the issues related to extrapolation in the biological and social sciences. But I do hope that the present work will be a useful point of departure for those interested in extrapolation and interconnected methodological and philosophical topics.

Appendix

A. CORRELATION AND THE PROBLEM OF DISJUNCTIVE CAUSAL FACTORS

As explained in section 2.3.1, the problem of disjunctive factors is an objection to the probability-raising definition of positive causal relevance. The essential point is that when X is a variable that may take more than two values, whether a particular value of X raises or lowers the probability of Y may depend upon the relative frequency of the other values of X in the population. Yet it seems that whether X promotes or inhibits Y should not depend on how frequently different values of X happen to occur. I argued in section 2.3.1 that this intuition makes sense, given the reasonable assumption that claims about positive relevance aim to provide information concerning the consequences of interventions on the cause. Humphreys's (1989, 40–41) example illustrates how the problem of disjunctive causal factors can arise with regard to the probability-raising definition of positive causal relevance. In fact, the problem of disjunctive factors would also arise if positive causal relevance were defined in terms of positive correlation. That is, whether X and Y are positively or negatively correlated can depend on the frequency with which particular values of X occur.

Recall that the correlation between X and Y is defined thus:

$$\rho(X,Y) = \frac{\text{cov}(X,Y)}{\sqrt{\text{var}(X)\ \text{var}(Y)}}$$

Hence, so long as the variances of X and Y are strictly greater than zero, whether the correlation is negative or positive is determined by the covariance. Consequently, it is necessary only to show that differences in the probabilities of distinct values of X can switch the covariance of X and Y from positive to negative. In particular, suppose that X and Y each have three values: 0, 1, and 2. Now consider these two joint distributions:

X	Y	p
2	2	.5
2	1	.05
2	0	.05
1	2	.003
1	1	.00001

X	Y	p
2	2	.005
2	1	.0005
2	0	.0005
1	2	.3
1	1	.001

1	0	.00299	1	0	.299
0	2	.15	0	2	.15
0	1	.15	0	1	.15
0	0	0.94	0	0	.094
	Distribution 1			Distribution 2	

These two distributions are identical except that in the first $P(X = 2) = .6$ and $P(X = 1) = .006$, while in the second $P(X = 2) = .006$ and $P(X = 1) = .6$. In other words, the two distributions differ only with regard to the probabilities of these two values of X. In both distributions, the values of X and Y tend to coincide when $X = 2$ but not when $X = 1$. As a result, X and Y are positively correlated in distribution 1 but negatively correlated in distribution 2.

This can be seen through calculating the covariance of X and Y in distributions 1 and 2. Recall that the covariance of X and Y, cov(X, Y), equals $E(XY)$—$E(X)E(Y)$. In distribution 1, we have:

$$E(X) = (2 \times .6) + .006 = 1.206$$
$$E(Y) = (2 \times .653) + .20001 = 1.50601$$
$$E(XY) = (4 \times .5) + (2 \times .05) + (2 \times .003) + .00001 = 2.10601$$
$$\text{cov}(X,Y) = 2.10601 - (1.206 \times 1.50601) = .28976194$$

Thus, X and Y are positively correlated in distribution 1. But in distribution 2 we have:

$$E(X) = (2 \times .006) + .6 = .612$$
$$E(Y) = (2 \times .455) + .1515 = 1.0615$$
$$E(XY) = (4 \times .005) + (2 \times .0005) + (2 \times .3) + .001 = .622$$
$$\text{cov}(X,Y) = .622 - (.612 \times 1.0615) = -.027638$$

In distribution 2, therefore, X and Y are negatively correlated.

This example illustrates that the problem of disjunctive causal factors that Humphreys and others raised as an objection to the probability-raising definition of causal relevance is also pertinent to correlation. That is, consider the claim that X is positively causally relevant to Y exactly if X is positively correlated with Y (in a context in which there is no confounding and in which Y cannot cause X). Just like the probability-raising definition, this proposal entails that whether X is positively or negatively causally relevant to Y can depend upon the probabilities of the various values of X. But as explained in section 2.3.1, from the perspective of a manipulationist account of causation, this is a highly undesirable characteristic.

B. QUANTITATIVE EXTRAPOLATION WITHOUT CONSONANCE

The extrapolation theorem presented in section 6.2.2 was limited insofar as it specified conditions only for extrapolating claims about positive or negative causal relevance and in presupposing consonance. Yet one might wish to extrapolate a quantitative claim about a causal effect, and one might also wish to extrapolate probabilistic causal claims in cases in which counteracting mechanisms may be present. Of course, quantitative extrapolation is easy when the base population is representative of the target, but it is often the case that this assumption is doubtful or known to be false. But even when the base population fails to be representative of the target, it may nevertheless be what I call *cell-representative*. The base population is cell-representative of the target when there is a partition of the base population into cells such that the strength of the causal effect within each cell in the base population is a good approximation of the strength of the effect in the corresponding cell in the target. A base population can be cell-representative without being representative if the relative frequencies of the cells differ between the two populations.

Elaborating this idea requires a measure of strength of causal effect. One commonly used measure of causal efficacy is the *mean difference*,[1] according to which the impact of X upon Y is given by $E(Y \mid do(x)) - E(Y \mid do(x_0)) =_{df} \Delta E(Y \mid do(x))$. Recall that x_0 is the comparative value of X, usually zero. For simplicity, I restrict attention to the special case in which the cause and effect are binary. In this case, the mean difference is $P(Y = 1 \mid do(X = 1)) - P(Y = 1 \mid do(X = 0)) =_{df} \Delta P$. A partition of a population consists of a mutually exclusive and collectively exhaustive collection of subsets of that population, in the simplest case, those who possess a particular property and those who do not. The cells of partitions will be numbered $1, 2, \ldots, n$. The probability function for the ith cell of the partition is represented by P_i. So, for example, ΔP_2 is $P_2(Y = 1 \mid do(X = 1)) - P_2(Y = 1 \mid do(X = 0))$. In this context, the population P′ is *cell-representative* of P with respect to X and Y, given the partition $i = 1, 2, \ldots, n$ just in case $\Delta P'_i \approx \Delta P_i$ for all i, where P' and P are the probability functions for the populations P′ and P, respectively. Thus, the idea is that strengths of the causal effects are approximately equal within the cells in the two populations, and that differences in the overall effect between the two populations result only from differences in the proportions of these cells.

Clearly, whether it is reasonable to assume that the base population is cell-representative of the target depends on the choice of partition. In section 6.2.1, it was presumed that the partition was by the particular set of undisrupted mechanisms possessed by the individual. If practically possible, this would be a promising way to partition, since differences in the strength of the causal effect presumably result from differences in mechanisms. However, it will generally be difficult, if not impossible, to accurately decide precisely which combination of mechanisms is present in a given individual. The presence or absence of detectable factors that promote or interfere with the mechanisms in question would provide some guidance for such purposes. In the case of the effect of HIV exposure upon AIDS, this would include such things as availability of anti-retroviral therapies or host resistance factors (such as the mutation affecting the R-5 co-receptor). But it is doubtful that the subgroups identified by such indicators will consist of individuals possessing the precisely same combinations of mechanisms.

This might seem like a serious problem, since equations (6.8) and (6.9) were derived on the assumption that one was partitioning by combinations of mechanisms. For instance, equation (6.9) told us that $\Delta P = \sum_{i=1}^{n} \varphi_i \Delta P_i$, where each i indicates a specific combination of mechanisms from X to Y. Fortunately, however, these equations can be derived for different partitions, so long as the properties by which one partitions are independent of the cause when it is set by an intervention. That is, if the cells in the partition are $i = 1, 2, \ldots, n$, then the key premise is that $P(i \mid do(x)) = P(i)$, for each i. Given the definition of an ideal intervention, this premise is reasonable so long as the properties by which one partitions are not effects of X. For example, possession of the mutation inhibiting the expression of the R-5 co-receptor is presumably not an effect of HIV exposure.

Consider how quantitative extrapolation on the basis of a cell-representative base population could work in the AFB_1 example. Susceptibility to the carcinogenic effects of AFB_1 is known to depend on exposure to the hepatitis B virus (HBV).[2] There is also evidence that heightened sensitivity to mutagens is also a co-factor (Wu et al. 1998), although the basis of these variations in AFB_1 susceptibility remains somewhat unclear (cf. McGlynn et al. 2003). It is likely due in part to HBV exposure (Sohn et al. 2000), but the importance of other factors, such as congenital genetic variations, is still uncertain. At present, then, HBV exposure is the most firmly established factor for susceptibility to AFB_1 carcinogenesis as well as something that can be measured reliably. Hence, given that it is likely that HBV exposure is not an effect of exposure to AFB_1, HBV would appear to be a good property by which to partition.

Imagine that one is interested in estimating the strength of the causal effect of AFB_1 on liver cancer in North America from data from China, where exposure to AFB_1 is more common and consequently where there are more extensive data sets. Thus, P′ and P in this case would be the populations of China and North America, respectively. Letting X represent exposure to AFB_1 and Y, occurrence of liver cancer, ΔP is the strength of the causal effect in the North American population, and similarly for $\Delta P'$. Since HBV is much more common in China than in North America, it is obvious that it would be unreasonable to regard $\Delta P'$ as a good estimate of ΔP. Nevertheless, the Chinese population might serve as an approximate guide to the North American one if we partition by those who have been exposed to HBV and those who have not. Labeling these two groups

1 and 2, respectively, $\Delta P_1 = P_1(Y = 1 \,|\, do(X = 1)) - P_1(Y = 1 \,|\, do(X = 0))$ is the strength of the causal effect among those in the North American population exposed to HBV, and similarly for ΔP_2. Let φ_1 and φ_2, respectively, be the relative frequencies of those exposed and not exposed to HBV. Then if AFB_1 is not a cause of HBV exposure, we can derive (as explained in section 6.2.1) the following equation.

$$\Delta P = \varphi_1 \Delta P_1 + \varphi_2 \Delta P_2 \tag{6.10}$$

Thus, if the Chinese population is approximately representative of the North American one with regard to the strength of the carcinogenic effect of AFB_1 among those exposed to HBV and those not exposed (i.e., $\Delta P_1'$ and $\Delta P_2'$ provide reasonably good estimates of ΔP_1 and ΔP_2, respectively), then the strength of the overall causal effect in the North American population, ΔP, can be computed, given the relative frequency of exposure to HBV in North America.[3]

Notice that the above reasoning does not depend on assuming consonance: if it is possible to estimate the strength of the causal effect in each cell of a cell-representative base population, then the overall effect in the target population can be estimated as explained above. For example, one could imagine a case like the above but in which ΔP_1 is positive and ΔP_2 is negative. However, consonance would be a useful assumption if it were possible to estimate the strength of the causal effect only in some cells and not others or if the base population were representative of the target for only some cells. For instance, suppose that in the aflatoxin example the strength of the causal effect could be estimated only for those who have been exposed to HBV. Then, given consonance, $\Delta P \geq \varphi_1 \Delta P_1$, which means that a lower bound can be placed on ΔP. This inference would not be valid, however, if consonance did not obtain, since in that case the second term on the right-hand side of (6.10) could be negative.

But what if only some cells of P′ are representative of those in P and consonance is not plausible? Informative extrapolations may be possible even in this unfavorable situation. Since the maximum and minimum values of ΔP are 1 and -1, it is possible to compute extreme upper and lower bounds from $\varphi_1 \Delta P_1$ when consonance is not assumed. That is, given that $\varphi_1 \Delta P_1$ is estimated, $\varphi_1 \Delta P_1 - (1 - \varphi_1) \leq \Delta P \leq \varphi_1 \Delta P_1 + (1 - \varphi_1)$. In other words, the lower bound is what results from the assumption that the strength of the effect is -1 in the remainder of the population (whose relative frequency is $1 - \varphi_1$), while the upper bound results from the assumption that the strength of the effect in the remainder of the population is 1. The breadth of the range contained within these upper and lower bounds obviously varies inversely with the size of φ_1, that is, the greater the proportion of the population for which a causal strength is estimated, the narrower the interval of possible values of ΔP. If one is primarily concerned to know whether the overall effect is positive or negative, the value of ΔP_1 also has a bearing on how informative the interval is, since for a given φ_1, the farther ΔP_1 is from zero, the more the interval is skewed to the positive or negative side of the scale. Of course, some areas within the interval may reasonably be judged to be more probable than others. For instance, the strength of the causal effect is 1 when the cause is both necessary and sufficient for the effect, and one might have good reason to think it extremely improbable that this would be the case.

C. D-SEPARATION

D-separation is a graphical concept whose interest lies in the following, highly nontrivial fact: for directed acyclic graphs, d-separation indicates exactly those conditional and marginal probabilistic independencies entailed by the causal Markov condition (CMC) (cf. Pearl 2000, 18).[4] A graph consists of a set of nodes, some or all of which are linked by lines, or edges. Typically, the nodes are understood to represent variables. A graph is said to be *directed* if each edge has an arrowhead at exactly one end. For example, consider the graphs in Figure A.1.

Graph (A) is directed, but (B) and (C) are not: (B), because of the undirected edge between W and N, and (C), because of the double-headed arrow between W and S. A graph is said to be *acyclic* if it does not contain any sequence of arrows all aligned head to tail that begin and end at the same node. For example, the graph in Figure A.2 contains a cycle.

In contrast, graph (A) in Figure A.1 is both directed and acyclic.

D-separation, then, is defined as follows:

> A path p is said to be *d-separated* (or *blocked*) by a set of nodes **S** if and only if
>
> 1. p contains a chain $i \rightarrow m \rightarrow j$ or a fork $i \leftarrow m \rightarrow j$ such that the middle node m is in **S**, or
> 2. p contains an inverted fork (or *collider*) $i \rightarrow m \leftarrow j$ such that the middle node m is not in **S** and such that no descendant of m is in **S**.
>
> A set **S** is said to *d-separate* X from Y if and only if **S** blocks every path from a node in X to a node in Y. (Pearl 2000, 16–17).
>
> X and Y are said to be *d-connected* by a set **S** if **S** does not d-separate them.

Notice that item 1 of the definition corresponds to the screening-off rule (see sections 4.4.1 and 9.1). Meanwhile, item 2 corresponds to the rule that

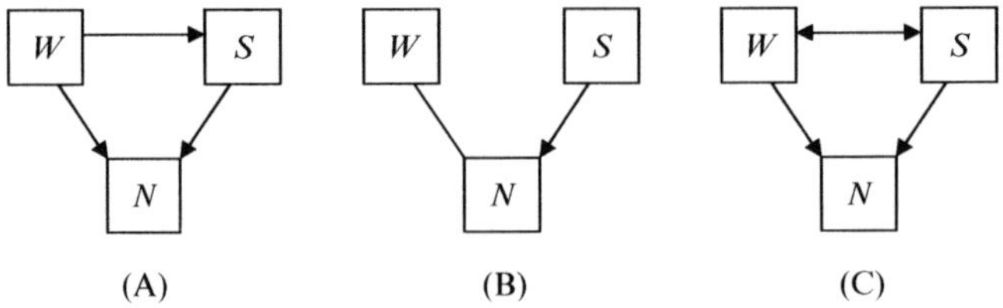

Figure A.1 Directed versus undirected graphs

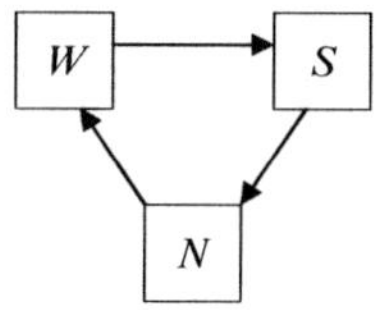

Figure A.2 A directed graph with a cycle

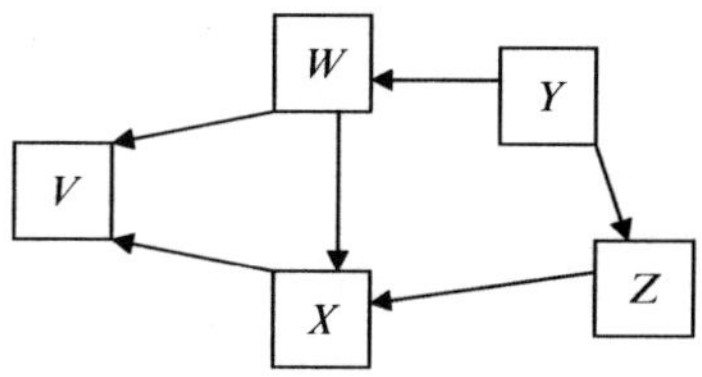

Figure A.3 An illustration of d-separation

conditioning on colliders (or their descendants) may induce probabilistic dependence (see Figure 9.2 in section 9.1).

As an illustration, consider the directed acyclic graph in Figure A.3. In this graph there are four paths between V and Z:

(1) $V \leftarrow W \leftarrow Y \rightarrow Z$
(2) $V \leftarrow W \rightarrow X \leftarrow Z$
(3) $V \leftarrow X \leftarrow W \leftarrow Y \rightarrow Z$
(4) $V \leftarrow X \leftarrow Z$.

Suppose that **S** is the empty set. Then paths (1), (3), and (4) are not blocked by **S,** though (2) is owing to the collider at X. Therefore, V and Z are d-connected by the empty set. Recall that the faithfulness condition (FC) asserts that the *only* probabilistic independence relations are those entailed by the CMC. Thus, if the graph satisfies the FC, V and Z are probabilistically dependent when no variables are conditioned upon (i.e., they are marginally dependent).

Suppose that $\mathbf{S} = \{X, Y\}$. Then paths (1), (3), and (4) are blocked by **S**. However, path (2) is not, since X is a collider on that path and X is **S**. Hence, V and Z are d-connected by $\{X, Y\}$. Thus, if the graph satisfies the FC, then V and Z are probabilistically dependent conditional on this set of variables.

Finally, suppose that $\mathbf{S} = \{W, X, Y\}$. In this case, all four paths are blocked by **S**. Thus, $\{W, X, Y\}$ d-separates V from Z. Consequently, if the graph in Figure A.3 satisfies the CMC, then V and Z are probabilistically independent conditional on $\{W, X, Y\}$.

D-separation also characterizes exactly those probabilistic independencies entailed by linear cyclic structures with independent error terms (Richardson and Spirtes [1999]). That result is of particular interest because d-separation and the CMC do not coincide for cyclic directed graphs. To see how this is so, consider the graph in Figure A.4.

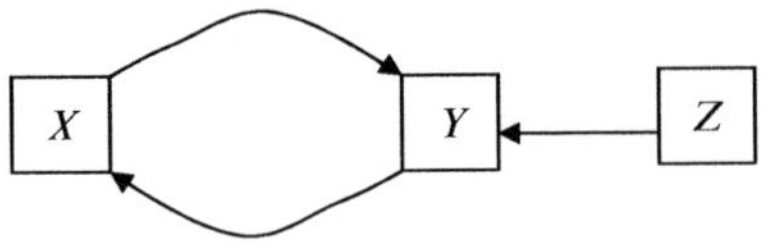

Figure A.4 How d-separation and the CMC differ for cyclic graphs

Note that Z is not a descendant of X, and the only parent of X is Y. Hence, if this graph satisfied the CMC, then X would be probabilistically independent of Z conditional on Y. However, {Y} does not d-separate X from Y, since Y is a collider on the path $X \rightarrow Y \leftarrow Z$. Given the proof that d-separation characterizes exactly those independence relationships entailed by linear cyclic structures with independent error terms, the natural conclusion is that d-separation is a more trustworthy guide for cyclic graphs than the CMC.

Notes

Notes to Chapter 1

1. The passage is from *Conjectures and Refutations* (1989 [1963], 72).
2. This is, of course, a classic problem in toxicology; cf. Calabrese (1991).
3. For example, see Manski and Garfinkel (1992).

Notes to Chapter 2

1. It is often assumed that causal structures are "causally complete" in the sense that no common causes of the variables have been omitted (Scheines 1997, 188). However, this assumption is not required for the account of extrapolation developed in this book.

2. See Woodward (2003, 98) for a similar definition.

3. However, see Menzies and Price (1993) for an attempt to transform a manipulationist account of causation into a reductive definition. See Hausman and Woodward (1999) and Dowe (2000) for a critique of this argument. See Woodward (2003, 20–22) for a defense of the claim that an account of causation can be illuminating without being a conceptual analysis (or, in his term, "reductive").

4. When X is a continuous variable, this function would be more naturally replaced by one indicating the probability distribution of Y conditional on an ideal intervention setting the value of X within some narrowly circumscribed interval, rather than to a specific value. However, I set aside this complication.

5. As is illustrated by such journal article titles as "The Estimation of Causal Effects from Observational Data" (Winship and Morgan 1999) and "Identification of Causal Effects Using Instrumental Variables" (Angrist, Imbens, and Rubin 1996).

6. Ellery Eells (1991, 24–25), who also interprets causal generalizations as being relative to populations, understands the notion of population in a rather different way than proposed here. For example, Eells requires that causal generalizations be relative not only to a flesh-and-blood population, but also to an abstract population type. I choose not to follow this approach, since I think it leads to unnecessary complications (cf. Eells 1991, 28–33). See Dupré (1993, 194–201) for a discussion of some the difficulties with Eells's approach to populations.

7. See Woodward (2003, 40) for a similar definition of causal relevance.

8. That is, the expected value of the product of X and Y minus the product of the expected value of X and the expected value of Y. The expected value of X equals $\Sigma xP(X = x)$.

9. See section A of the Appendix for a numerical illustration of this point.

10. Humphreys's proposal differs from Holland's in requiring that there be an objectively correct neutral comparison term. As Hitchcock (1993, 344–45) points out, there are difficulties with this proposal.

11. This intuition is shared by others as well (cf. Selten 2001, 31).

12. For example, for each individual in the population, define the variable *E* such that the closer the individual's wealth to the mean, the higher the value of *E*. Then increases in the redistribution of wealth from rich to poor yield increases in the expected value of *E*.

13. Note that it is possible for *X* to be both comparative and monotonic neutral with respect to *Y* within an interval θ while being causally relevant to *Y*. This could occur in two ways: changes of *X* might alter the distribution of *Y* only outside of the interval θ, or changes in *X* might alter the distribution of *Y* without changing its expected value. Note that neither of these two scenarios is possible if both *X* and *Y* are binary variables. This point will turn out to be surprisingly important to the discussion of extrapolation in Chapter 6.

14. This is an example of what is known as "hormesis." Similar response patterns occur widely in toxicology (cf. Calabrese and Baldwin 1999).

15. Observe that this condition entails that x_0 is not in θ, which is the one universal constraint on the choice of x_0 that I would insist on.

16. Brian Skyrms (1980, 108–9) suggests a weaker version of this requirement, which demands only that a positive causal factor not lower the probability in any subpopulation. Skyrms's concept is similar to consonance, discussed in Chapter 6.

17. Examples of this sort are examined in more detail in section B of the Appendix.

Notes to Chapter 3

1. See Hausman (1998, 13–17) for a critique of the conserved quantity theory on the grounds that it fails to distinguish causally relevant and irrelevant interactions and fails to account for causal asymmetries.

2. In Carnap's (1936) version of this account, the antecedently understood terms were presumed to be drawn from an "observation language." I follow Lewis (1970) and others (cf. Papineau 1996) in rejecting this requirement.

3. The transfer theory is, together with Salmon's (1984) proposal, one of the ancestral sources of Dowe's position. It is presented in Aronson (1971) and in Fair (1979). The transfer theory differs in some important respects from the conserved quantity theory, and unfortunately Dowe does not elaborate on how the Ramsey-Lewis approach would work in the case of his own theory.

4. For instance, Hitchcock (2003) advocates a pluralistic approach to causation.

5. Probably the most important graphical concept for this purpose is d-separation (cf. Pearl 2000, 16–20).

6. The term "Bayesian network" derives from the original (and continuing) use of directed graphs and probability distributions to implement expert learning and judgment in artificial intelligence (cf. Pearl 1988). In the context of causal inference, the name does not indicate a commitment to a Bayesian methodology.

7. In Simon's original parable, Hora and Tempus are interrupted by telephone calls, so that Tempus must continually restart construction from scratch, while Hora need only restart the last module. As Watson and Pollack (2005, 448) point out, it is difficult to interpret the original parable as an example of how

modularity enhances evolvability, since it does not involve a search through a space of possibilities for a solution to a problem.

8. A number of further empirical case studies concerning modularity and evolution can be found in Schlosser and Wagner (2004).

Notes to Chapter 4

1. The T-helper count in a healthy person is usually between 900 and 1200 per microliter.

2. M-tropic HIV can also infect T-helper cells circulating freely in the bloodstream, but not those present in lymphoid tissue. This latter category constitutes the vast majority of T-helper cells (Stine 2000, 129, 141).

3. HIV can also destroy T-helper cells in several other ways (cf. Kalichman 1998, 20).

4. SGS note this limitation of the standard, directed graph formalism (2000, 24–25). Geiger and Heckerman (1991) suggest a rather different device for representing interactions via causal graphs than that developed here.

5. For more on this topic, see Pearl (2000), SGS (2000), Shipley (2000), Rosenbaum (2002), and Neopolitan (2004). For commentary, see McKim and Turner (1997) and Glymour and Cooper (1999).

6. The first published account of the isolation of the HIV virus is Barre-Sinoussi et al. (1983).

7. T-helper cells circulating in the bloodstream typically express both the R5 and X4 co-receptors, and hence are susceptible to infection by both strains (Stine 2000, 141). However, the vast majority of lymphocytes occur in lymphoid tissue (ibid., 129).

8. M-tropic and T-tropic HIV are also often distinguished on the grounds that the latter, but not the former, produce syncytia (multinucleate masses of protoplasm not separated into distinct cells). Thus, M-tropic HIV is often labeled NSI (nonsyncytium-inducing) and T-tropic HIV, SI (syncytium-inducing).

9. For example, see the article "Immune to a Plague: Gene for Immunity to AIDS Discovered," in *Discover* magazine (Radetsky 1997). This article also provides a lively recounting of the path to the discovery of the thirty-two-base pair deletion in the gene for the R5 co-receptor.

10. First articulated, in a somewhat different form, by Hans Reichenbach (1956, 157).

11. This proposition can be extended to cyclic systems, at least for linear models, by employing a generalized version of the CMC (cf. Spirtes 1995; Koster 1996; Richardson and Spirtes 1999; SGS 2000, 297–99).

12. It is an interesting question whether, in the absence of randomization, an intervention might be exogenous, yet not probabilistically independent of other exogenous variables. Some have argued that such a thing occurs for pairs of causally unrelated variables both of which exhibit a time trend (cf. Sober 2001). Hoover (2003) argues that such cases are not in fact genuine counterexamples to the PCC, while Steel (2003, 316) points out that such problems cannot arise in randomized controlled experiments.

13. The same point is made by Scheines et al. (1998, 171–73) in response to Woodward (1998, 129–35).

14. See their theorem 3.2 (2000, 41–42) and its proof (2000, 383–84).

15. In mathematical jargon, **L** says that any subset of the space of parameterizations of Lebesgue measure zero has probability zero. The "**L**" tag of the assumption is for "Lebesgue measure."

16. Glymour (1999, 161) responds to Cartwright by claiming that reliable causal inference is impossible without the FC. Even if this claim were true, it would not follow that the objection is mistaken, but only that the reliable causal inference is more narrowly restricted than one might have hoped.

17. One might also ask whether SGS's theorem extends to cases in which causal relationships are nonlinear. SGS conjecture that it does (2000, 42), and Meek (1995) shows that the theorem holds for causal models with discrete variables.

18. That it would be unreasonable to insist that sets of Lebesgue measure zero must always have probability zero is noted by Pearl (1998, 121). SGS also acknowledge the point (2000, 66) but do not explain why sets of Lebesgue measure zero should have zero probability in the sorts of cases relevant to their theorem.

19. This conclusion bears some similarity to Pearl's (1998, 121) and Woodward's (1998, 142–47) suggestion that the FC is a reasonable assumption when parameters vary independently of one another while causal structure remains constant. For more detailed examination of this matter, see Steel (2006).

20. A consequence of this point is that examples of relatively simple technological devices in which near violations of the FC can be made probable do not show that near exceptions to the FC are likely more generally. In regard to this, see Cartwright's "solition" example, which she uses to motivate her objection to the FC (1999, 30–31, 118).

21. Chu et al. (2003) demonstrate an obstacle to the screening-off rule (e.g., variables related only as effects of a common cause C are independent conditional on C) in studies that aim to infer gene regulatory networks from microarray data. However, as they observe (Chu et al. 2003, 1147), this difficulty is not relevant to gene knockout experiments, which are my concern here.

22. For additional cases, see Liljegren et al. (2000) and Kurihara et al. (2001).

23. The same point is also noted in Tymms and Kola (2001, 5–6).

24. This section is titled "Mammalian Genetic Models with Minimal or Complex Phenotypes."

25. See Joyner (2000, chapters 3, 4, and 5, for a detailed description of such procedures.

Notes to Chapter 5

1. Although the term "extrapolation" suggests a situation in which there is no overlap between the target and base populations, the proposals advanced in this book are pertinent to cases in which the two populations are not disjoint. For example, the target population might be a proper subset of the base population, as in a medical example in which a physician wishes to judge whether a treatment that is effective with regard to the general population is also effective for some subgroup. Thus, the discussion in this chapter is relevant to all three of the examples of the problem of extrapolation in heterogeneous populations listed at the start of Chapter 1.

2. I thank Jim Woodward (personal communication) for suggesting this concise formulation of the issue.

3. See especially Calabrese (1991) and Hengstler et al. (1999).

4. They reiterate this criterion of CAM-hood in several places (cf. 1996, 113; 1993a, 122; 1993b, 326). LaFollette and Shanks's definition of a CAM is also adopted by Ankeny (2001, S256) and Guala (2005, 199).

5. I owe this objection to LaFollette and Shanks' definition of a CAM to Megan Delehanty.

6. They also claim—not very reasonably, in my judgment—that the *opposite* is accepted opinion among scientists conducting animal research (1993a, 119). See Hengstler et al. (1999, 919) for a clear statement that evidence is needed to establish the appropriateness of an animal model. Likewise, see Schaffner (2001) for a description of debates among scientists regarding the appropriateness of several animal models for specific extrapolations.

7. LaFollette and Shanks reiterate the extrapolator's circle at various points in their book (1996, 23, 27, 169).

8. LaFollette and Shanks cite Hempel (1965, 441). For other classic statements of distinction, see Popper (2002 [1959], 7–8) and Reichenbach (1938, 6–7).

9. They coin the term ''modeler's functional fallacy'' to refer to the belief that similarity in function entails similarity of mechanism.

10. I thank Fred Gifford for pointing out this ethical implication of my account of extrapolation.

Notes to Chapter 6

1. Mitchell (2002a) also argues that generalizations of the biological sciences do not follow the pattern of exclusive cp laws.

2. See Earman and Roberts (1999) and Earman, Roberts, and Smith (2002) for arguments that cp laws, owing to their open-ended escape clauses, can serve no legitimate scientific purpose. For an example of an economist dismissing such arguments as a ''foolish'' case of throwing the baby out with the bathwater, see (Persky 1990, 192–93).

3. Spohn's (2002) approach to cp laws in terms of ranking functions is similar to both Lange's and Schurz's proposals.

4. I borrow this label from Woodward (2002b).

5. Morreau (1999, 164) observes that it is a common problem for the completer approach that it allows both a sentence and its contrary to count as cp laws.

6. See Earman and Roberts (1999) and Schurz (2001b, 2002) for critiques of these versions of the completer approach. For criticisms of Fodor's proposal, see Schiffer (1991) and Mott (1992).

7. I do not explore the extrapolation of quantitative probabilistic causal claims in this chapter. See section B of the Appendix for some preliminary exploration of this topic.

8. The same concern is explored in Gold et al. (1992), Calabrese and Baldwin (1999), and Hengstler et al. (2003).

9. Lipton (1999) and Kincaid (1996, 63–70) offer accounts of cp laws that take their inspiration from Cartwright. See Smith (2002) for an argument that physical examples such as the law of universal gravitation in fact provide little support for Cartwright's analysis. See Cartwright (2002a) for a defense of her interpretation of cp laws.

10. The classic text in this genre is Kelly (1996).

11. This is a reformulation of Glymour's proposal, which appears to contain a typographical mistake. He writes: ''The learner verifies that *normally X* for a data sequence if only a finite number of these conjectures are in error or is of the form *if*

A_n *then* $\sim X$, and falsifies that *normally* X for a data sequence if only a finite number of these conjectures are in error or is of the form *if* A_n *then* X'' (2002, 401). The verification case is equivalent to my formulation. But the falsification case is rather odd, since it entails that a learner could avoid falsifying *normally* X by issuing infinitely many erroneous conjectures. Yet an infinite sequence consisting solely of incorrect conjectures would seem to be as clear a case of falsifying *normally* X as one could imagine. My guess is that the passage was intended to say that the learner ''falsifies that *normally* X for a data sequence if *an infinite* number of these conjectures are in error or of the form *if* A_n *then* $\sim X$.'' This makes falsifying *normally* X equivalent to not verifying it.

12. That grand theoretical unification is an inappropriate ideal for biological sciences is one of the main themes of Mitchell (2003).

13. Since Lange interprets laws in general as inference rules, one might also attribute this insight to him. However, his defense of cp laws is primarily based on the Wittgensteinian notion that the meaning of a phrase (e.g., ''nothing interferes'') can be implicit in practice and need not depend on explicit necessary and sufficient conditions for its application (cf. Lange 1993, 2002).

Notes to Chapter 7

1. This insight was expressed by William Wimsatt (1976).

2. I borrow this delightful turn of phrase from Sterelny and Griffiths (1999, 149).

3. The expression ''gory details'' is taken from a memorable line in Kitcher (1984, 370).

4. The authors cite research suggesting that these premises are correct (Callaway et al. 1999, 2525).

5. Callaway et al. note that this implication of their model is supported by empirical data (1999, 2528). Recall that M-tropic strains can infect circulating T-cells (which display the R5 co-receptor) but not those residing in lymphatic tissue.

6. The extent of the difference and whether it will persist have been questioned, however (Cilliers et al. 2004).

7. Nagel (1979, 99), Wimsatt (1979, 352), and Rosenberg (2001, 157) cite correction as a goal of reduction.

8. My use of the term ''theory'' is intended to presuppose no specific analysis of what theories are (cf. Suppe 1974).

9. See Fodor (1975, 10–12), Kincaid (1990, 576), and Dupré (1993, 88) for succinct presentations of the layer-cake model of reduction.

10. For example, see Rosenberg (2001, 136).

11. See Schaffner (1993a, 328) for an acknowledgment of this point. Sarkar also emphasizes that synthetic identities were never a requirement in Nagel's model of reduction (cf. 1998, 25).

12. Schaffner also rejects requirement 3 as a desideratum of reductions (cf. 1993a, 340).

13. Correction would seem to be precluded, since the layer-cake model presumes that the higher-level theory is deduced from the lower-level one (so the higher-level theory must be true if the lower-level one is). However, correction could be brought into the layer-cake picture if one supposed that what is deduced is not precisely the higher-level theory, but some modified, corrected version of it (cf. Schaffner 1967).

14. Not even Oppenheim and Putnam claimed that the layer-cake model represented the *only* possible type of reduction (cf. 1958, 8). Rather, they advanced the layer-cake model as an account of reduction capable of explicating the intuitive notion of the unity of science.

15. Sarkar rejects synthetic identities as a necessary condition for biological reductions on these grounds (1998, 36, 62).

16. My use of the term "level" is similar to Sarkar's use of "realm" (cf. Sarkar 1998, 39–47).

17. See Delehanty (2005) for a defense of token-token reductionism against the "context objection."

18. Kitcher includes two additional claims on this list: (4) states that the representations accepted by science at any given time may not all be mutually consistent, while (3) asserts that any such inconsistencies are the result of the imperfections of the current state of science (2002, 570). There is a minor controversy between Kitcher and Longino regarding (3) (cf. Longino 2002a, 184; 2002b, 575–76; Kitcher 2002, 571), but that disagreement is immaterial to our concerns here.

19. See Kim (1999) and Delehanty (2005) for critiques, and Humphreys (1997) for a defense of strongly emergent properties.

20. Likewise, see his statement that higher-level properties are not "causally inert" (1993, 101).

21. See Steel (2004, 69).

22. Rosenberg also identifies macromolecules as the fundamental level of biological description (2001, 162).

23. For example, Cartwright's pluralism is more overtly ontological than Kitcher's, and she makes a point of criticizing token-token reductionism (cf. 1999, 32–33). For a critique of Cartwright's pluralism from a perspective that is sympathetic to Kitcher's, see Ruphy (2003).

Notes to Chapter 8

1. Of course, the task is made easier in this case by the fact that it was possible to choose which mechanism to implement: the robustness of a mechanism is one factor in favor of selecting it.

2. This example suggests that structure-altering interventions directly affect more than one variable, and hence violate item (b) in definition 2.1.

3. This is a way to interpret the main thrust of the Lucas critique (cf. Woodward 2000, 220–21).

4. Of course, additional assumptions about preferences, such as transitivity and completeness, would typically be made. See Hausman (1992, chap. 1) for an accessible discussion of the standard conditions that preferences are assumed to satisfy in rational choice models.

5. For instance, this specification test assumes that groups evenly matched at the earlier time will continue to be evenly matched the later time (cf. Heckman and Hotz 1989, 666).

6. These authors also refer to "selection bias"—the existence of common causes of program participation and the outcome of interest, say, earnings—as an extrapolation problem (Manski and Garfinkel 1992, 13). Yet selection bias is a challenge for estimating a causal effect in a given context, and not a problem having to do with extrapolating a causal effect from one population and context to others.

7. For assessments of the effects of the 1996 welfare reform on the incomes of former recipients, see Danziger et al. (2002), Wolfe (2002), DeParle (2004, chap. 17), Robbins and Barcus (2004), and Ozawa and Yoon (2005).

Notes to Chapter 9

1. See Pearl (2000, 17) and SGS (2000, 24–25) for further discussion of this type of example.

2. See Angrist, Imbens, and Rubin (1996) and Rosenbaum (2002, 180–88) for a discussion of details.

3. See Little (1991, 25).

4. Kincaid attributes this claim to Elster (cf. Elster 1989, 4) and to Little (cf. Little 1991, 25).

5. A very similar argument is found in Papineau (1978, 54).

6. See Spirtes, Glymour, and Scheines (2000, chap. 4) for a thorough discussion and more complex examples of statistically indistinguishable causal graphs.

7. There is a basis for ruling out the possibility that *X* is a cause of *Y* that is frequently appealed to in social research, namely, that *Y* is temporally prior to *X*. However, this reasoning depends only on the principle that an effect cannot precede its cause in time, which one might maintain independently of any convictions regarding mechanisms. Of course, that *Y* is prior in time to *X* does not rule out the possibility of common causes of *Y* and *X*.

8. Little reiterates the same position in more recent writings (cf. 1998, 213–14).

9. Danks (2005) gives an interesting normative proposal for how, given the CMC and FC, conclusions about the causal relationships among distinct yet related sets of variables can be integrated.

10. See section 3.5.1.

11. Since Ferguson's account of Yanomami warfare is so different from the popular perception of the topic—according to which the supposed incessant violence of the Yanomami is a grim portrait of our primitive ancestors—some background comments are in order. The popular view of the Yanomami is primarily due to Napoleon Chagnon's famous depiction of them as the "fierce people" (1968, 1974, 1988). Most lay people, I think, would be surprised to learn that nearly every anthropologist who has seriously studied the Yanomami rejects Chagnon's portrayal of them (see Sponsel 1998 for a good literature review).

12. I located eight reviews of Ferguson's book. Four are positive (Rivière 1996; Chernella 1997; Pollock 1997; Harris 1996). One of these positive reviews was written by Marvin Harris, a longtime champion of ecological explanations of Yanomami warfare (1977, 1984). In his review, Harris abandons his ecological hypothesis in favor of Ferguson's—at least as far as the Yanomami are concerned (Harris 1996, 416). A fifth review is generally positive in tone, but offers no clear verdict of approval or disapproval (Heinen and Illius 1996). One review is mixed, acknowledging that Ferguson had made a major contribution to the issue and granting that he had shown that conflicts regarding steel tools have been an important cause of Yanomami warfare (Colchester 1996). Nevertheless, this reviewer remained skeptical about the importance of this cause in comparison to others. The only negative review by an anthropologist that I found was Chagnon's (1996). I also found one very negative, brief, and sarcastic review written by a historian (Bellesiles 1998). In sum, aside from Chagnon, the reaction to Ferguson's book among anthropologists has been mostly positive. Moreover, Yanomami

specialists generally appear to regard his book as making a significant contribution that should reshape anthropological discussions of warfare. For example, in a review of the Yanomami warfare literature, Leslie Sponsel states that Ferguson's "work should force anthropologists to reevaluate previous ethnographies as well as to evaluate and design future research in light of the distinct *possibility* that what was formerly believed to be chronic, endemic 'primitive' or tribal warfare may actually have been triggered (or at least intensified) and transformed by contact (indirect or direct) with Western 'civilization' " (1998, 110).

13. Thus, I follow many current philosophers of science in rejecting the sharp distinction between the context of discovery and the context of justification drawn by logical empiricists. See section 5.4.3 for further discussion of this topic.

Notes to the Appendix

1. For example, the mean difference is often used as a measure of treatment impact in randomized controlled experiments. That is illustrated by the experimental evaluations of welfare-to-work programs discussed in section 8.2. However, other measures exist (for example, ratios) and may be preferable for some purposes.

2. See Kew (2003) for a good literature review on this topic.

3. This procedure bears an obvious similarity to stratification in observational studies (cf. Rosenbaum 2002, 77–82). However, there is an important difference, since stratification is a method for estimating a causal effect in a population from *statistical data* concerning that *same* population. In contrast, the inference of concern here is an extrapolation: given the *causal effect* in one population, one wishes to draw conclusions about the effect in *another* population.

4. The CMC was discussed in sections 4.4.1 and 9.1.

References

Abebe, Almaz, Dereje Demissie, Jaap Goudsmit, Margreet Brouwer, Carla L. Kuiken; Georgios Pollakis, Hanneke Schuitemaker, Arnaud L. Fontanet, and Tobias F. Rinke de Wit (1999). "HIV-1 Subtype C Syncitium- and Non-syncytium-inducing Phenotypes and Coreceptor Usage Among Ethiopian Patients with AIDS." *AIDS* 13: 1305–11.

Aldrich, Howard (1999). *Organizations Evolving*. London: Sage.

Alexandrova, Anna (2006). "Connecting Economic Models to the Real World: Game Theory and the FCC Spectrum Auctions." *Philosophy of the Social Sciences* 36: 173–92.

Ames, Bruce, Renae Magaw, and Lois Gold (1987). "Ranking Possible Carcinogenic Hazards." *Science* 236 (4799): 271–80.

Ancel, Lauren, and Walter Fontana (2000. "Plasticity, Evolvability, and Modularity in RNA." *Journal of Experimental Zoology (Molecular and Developmental Evolution)* 288: 242–83.

Angrist, Joshua (1990). "Lifetime Earnings and the Vietnam Era Draft Lottery: Evidence from Social Security Administrative Records." *American Economic Review* 80: 313–35.

Angrist, Joshua, Guido Imbens, and Donald Rubin (1996). "Identification of Causal Effects Using Instrumental Variables." *Journal of the American Statistical Association* 91: 444–55.

Angrist, Joshua, and Alan Krueger (1991). "Does Compulsory School Attendance Affect Schooling and Earnings?" *Quarterly Journal of Economics* 106: 979–1014.

——— (1992). "The Effect of Age at School Entry on Educational Attainment: An Application of Instrumental Variables with Moments from Two Samples." *Journal of the American Statistical Association* 87: 328–36.

Ankeny, Rachel (2001). "Model Organisms as Models: Understanding the 'Lingua Franca' of the Human Genome Project." *Philosophy of Science* 68: S251–61.

Aoki, Masahiko, and H. Takizawa (2002), "Information, Incentives, and Option Value: The Silicon Valley Model", *Journal of Comparative Economics* 30: 759–86.

Aronson, Jerrold (1971). "The Legacy of Hume's Analysis of Causation." *Studies in the History and Philosophy of Science* 2: 135–56.

Barre-Sinoussi, F., J. Chermann, F. Rey, M. Nugeyre, S. Chamaret, J. Gruest, C. Dauguet, C. Axler-Blin, F. Vezinet-Brun, C. Rouzioux, W. Rozenbaum, and L. Montagnier (1983). "Isolation of a T-Lymphotropic Retrovirus from a Patient at Risk for Acquired Immune Deficiency Syndrome (AIDS)." *Science* 220: 868–71.

Bartik, Timothy (2000). "Displacement and Wage Effects of Welfare Reform." In Card and Blank (eds.), 72–122.

Batterman, Robert (2000). "Multiple Realizability and Universality." *British Journal for the Philosophy of Science* 51: 115–45.

Batterman, Robert (2002). *The Devil in the Details: Asymptotic Reasoning in Explanation, Reduction, and Emergence*. Oxford: Oxford University Press.

Bechtel, William, and Jennifer Mundale (eds.) (1999). "Multiple Realizability Revisited: Linking Cognitive and Neural States." *Philosophy of Science* 66: 175–207.

Bechtel, William, and Robert Richardson (1993). *Discovering Complexity: Decomposition and Localization as Strategies in Scientific Research*. Princeton, NJ: Princeton University Press.

Beldade, Patrícia, Kees Koops, and Paul M. Brakefield (2002). "Modularity, Individuality and Evo-devo in Butterfly Wings." *Proceedings of the National Academy of Sciences* 99: 14262–67.

Bellesiles, Michael (1998). "Yanomami Warfare: A Political History." *Journal of the West* 37: 101–2.

Biti, Robyn, Rosemary French, Judy Young, Bruce Bennetts, Graeme Stewart, and Tong Liang (1997). "HIV-1 Infection in an Individual Homozygous for the CCR5 Deletion Allele." *Nature Medicine* 3: 252–53.

Block, Ned (1997). "Anti-Reductionism Slaps Back." *Philosophical Perspectives* 11: 107–32.

Bontly, Thomas (2006). "What Is an Empirical Analysis of Causation?" *Synthese* 151: 177–200.

Boyd, Richard, and Peter Richerson (2005). *The Origin and Evolution of Cultures*. Oxford: Oxford University Press.

Braithwaite, R. B. (ed.) (1954). *The Foundations of Mathematics and Other Logical Essays* by F. P. Ramsey. London: Routledge and Keegan Paul.

Brown, J. Brian, and Daniel Lichter (2004). "Poverty, Welfare, and the Livelihood Strategies of Nonmetropolitan Single Mothers." *Rural Sociology* 69: 282–301.

Burian, Richard (1993). "How the Choice of Experimental Organism Matters: Epistemological Reflections on an Aspect of Biological Practice." *Journal of the History of Biology* 26: 351–67.

Calabrese, Edward (1991). *Principles of Animal Extrapolation*. Chelsea, MI: Lewis Publishers.

Calabrese, Edward, and Linda Baldwin (1999). "Reevaluation of the Fundamental Dose-Response Relationship." *BioScience* 49 (9): 725–32.

Callaway, Duncan, Ruy Ribeiro, and Martin Nowak (1999). "Virus Phenotype Switching and Disease Progression in HIV-1 Infection." *Proceedings of the Royal Society of London* 266: 2523–30.

Camerer, Colin (1995). "Individual Decision Making." In Kagel and Roth (eds.), pp. 587–703.

Card, David, and Rebecca Blank (eds.) (2000). *Finding Jobs: Work and Welfare Reform*. New York: Russell Sage Foundation.

Carnap, Rudolf (1936). "Testability and Meaning." *Philosophy of Science* 3: 419–71; *Philosophy of Science* 4: 1–40.

Carroll, Glenn, and Anand Swaminathan (2000). "Why the Microbrewery Movement? Organizational Dynamics of Resource Partitioning in the U.S. Brewing Industry." *American Journal of Sociology* 106: 715–62.

Cartwright, Nancy (1983). *How the Laws of Physics Lie*. New York: Oxford University Press.

——— (1989). *Nature's Capacities and Their Measurement*. Oxford: Oxford University Press.

——— (1992). "Aristotelian Natures in the Modern Experimental Method." In Earman (ed.).
——— (1995a). "Causal Structures in Econometrics." In D. Little (ed.), *On the Reliability of Economic Models*. Boston: Kluwer Academic Publishers, pp. 63–74.
——— (1995b). "Reply to Eells, Humphreys and Morrison." *Philosophy and Phenomenological Research* 55: 177–87.
——— (1995c). "*Ceteris Paribus* Laws and Socio-Economic Machines." *The Monist* 78: 276–94.
——— (1999). *The Dappled World: A Study of the Boundaries of Science*. Cambridge: Cambridge University Press.
——— (2002a). "In Favor of Laws That Are Not Ceteris Paribus After All." *Erkenntnis* 57: 425–39.
——— (2002b). "Against Modularity, the Causal Markov Condition and Any Link Between the Two: Comments on Hausman and Woodward." *British Journal for the Philosophy of Science* 53: 411–53.
Castle, David (2001). "A Gradualist Theory of Discovery in Ecology." *Biology and Philosophy* 16: 547–71.
Cecilia, D., S. Kulkarni, S. Tripathy, R. Gangakhedkar, R. Paranjape, and D. Gadkari (2000). "Absence of Coreceptor Switch with Disease Progression in Human Immunodeficiency Virus Infections in India." *Virology* 271: 253–31.
Chagnon, Napoleon (1968). *Yanomamö: The Fierce People*. New York: Holt, Rinehart and Winston.
——— (1974). *Studying the Yanomamö*. New York: Holt, Rinehart and Winston.
——— (1988). "Life Histories, Blood Revenge, and Warfare in a Tribal Population." *Science* 239: 985–92.
——— (1996). "Yanomami Warfare: A Political History—Ferguson, R. B." *American Anthropologist* 98: 670–72.
Chai, Xiaoyong (2005). "Cognitive Preference Reversal and Market Price Reversal." *Kyklos* 58: 177–94.
Cheng, Patricia (1997). "From Covariation to Causation: A Causal Power Theory." *Psychological Review* 104: 367–405.
——— (2000). "Causality in the Mind: Estimating Contextual and Conjunctive Power." In Frank C. Keil and Robert A. Wilson (eds.), *Explanation and Cognition*. Cambridge, MA: MIT Press, pp. 227–54.
Chernella, Janet (1997). "Yanomami Warfare: A Political History—Ferguson, R. B." *American Ethnologist* 24: 227–29.
Chipman, Ariel (2002). "Variation, Plasticity and Modularity in Anuran Development." *Zoology* 105: 97–104.
Chu, Tianjiao, Clark Glymour, Richard Scheines, and Peter Spirtes (2003). "A Statistical Problem for Inference to Regulatory Structure from Associations of Gene Expression Measurement with Microarrays." *Bioinformatics* 19: 1147–52.
Cilliers, Tonie, Jabulani Nhlapo, Mia Coetzer, Dragana Orlovic, Thomas Keta, William Olson, John More, Alexandra Trkola, and Lynn Morris (2003). "The CCR5 and CXCR4 Coreceptors Are Both Used by Human Immunodeficiency Virus Type 1 Primary Isolates from Subtype C." *Journal of Virology* 77: 4449–56.
Colchester, Marcus (1996). "Yanomami Warfare: A Political History—Ferguson, R. B." *Journal of the Royal Anthropological Institute* 2: 549–50.
Connor, Ruth, and David Ho (1994). "Human Immunodeficiency Virus Type 1 Variants with Increased Replicative Capacity Develop During the Asymptomatic State Before Disease Progression." *Journal of Virology* 68: 4400–8.

Cowen, Tyler (1998). "Do Economists Use Social Mechanisms to Explain?" In Hedström and Swedberg (eds.), pp. 125–46.

Craver, Carl, and Lindley Darden (2001). "Discovering Mechanisms in Neurobiology: The Case of Spatial Memory." In Machamer, Grush, and McLaughlin (eds.), pp. 112–37.

Cubitt, Robin, Alistair Munro, and Chris Starmer (2004). "Testing Explanations of Preference Reversal." *Economic Journal* 114: 709–26.

Danks, David (2005). "Scientific Coherence and the Fusion of Experimental Results." *British Journal for the Philosophy of Science* 56: 791–808.

Danziger, Sheldon, Colleen Heflin, Mary Corcoran, Elizabeth Oltmans, and Hui-Chen Wang (2002). "Does It Pay to Move from Welfare to Work?" *Journal of Policy Analysis and Management* 21: 671–92.

Darden, Lindley (1991). *Theory Change in Science: Strategies from Mendelian Genetics.* New York: Oxford University Press.

——— (2002). "Strategies for Discovering Mechanisms: Schema Instantiation, Modular Subassembly, Forward/Backward Chaining." *Philosophy of Science* 69 (supplement): S354–65.

Darden, Lindley, and Carl Craver (2002). "Strategies in the Interfiled Discovery of the Mechanism of Protein Synthesis." *Studies in History and Philosophy of Biological and Biomedical Science* 33: 1–28.

Davies, Todd (1988). "Determination, Uniformity, and Relevance: Normative Criteria for Generalization and Reasoning by Analogy." In David Helman (ed.), *Analogical Reasoning: Perspectives of Artificial Intelligence, Cognitive Science, and Philosophy.* Boston: Kluwer Academic, pp. 227–50.

Dean, Michael, Mary Carrington, Cheryl Winkler, Gavin A. Huttley, Michael W. Smith, Rando Allikmets, James J. Goedert, Susan P. Buchbinder, Eric Vittinghoff, Edward Gomperts, Sharyne Donfield, David Vlahov, Richard Kaslow, Alfred Saah, Charles Rinaldo, Roger Detels, and Stephen J. O'Brien (1996). "Genetic Restriction of HIV Infection and Progression to AIDS by a Deletion Allele of the CKR5 Structural Gene." *Science* 273: 1856–62.

Delehanty, Megan (2005). "Emergent Properties and the Context Objection to Reduction." *Biology and Philosophy* 20: 715–34.

Deng, Hong Kui, Rong Liu, Wilfried Ellmeier, Sunny Choe, Derya Unutmaz, Michael Burkhart, Paola Di Marzio, Shoshana Marmon, Richard E. Sutton, C. Mark Hill, Craig B. Davis, Stephen C. Peiper, Thomas J. Schall, Dan R. Littman, and Nathaniel R. Landau (1996). "Identification of a Major Co-receptor for Primary Isolates of HIV-1." *Nature* 381: 661–66.

DeParle, Jason (2004). *American Dream: Three Women, Ten Kids, and a Nation's Drive to End Welfare.* London: Penguin Books.

Donohue, John, and Steven Levitt (2001). "The Impact of Legalized Abortion on Crime." *Quarterly Journal of Economics* 116: 379–420.

Dowe, Phil (2000). *Physical Causation.* Cambridge: Cambridge University Press.

Dragic, Tatjana, Virginia Litwin, Graham P. Allaway, Scott R. Martin, Yaoxing Huang, Kirsten A. Nagashima, Charmagne Cayanan, Paul J. Maddon, Richard A. Koup, John P. Moore, and William A. Paxton (1996). "HIV-1 Entry into CD4+ Cells Is Mediated by the Chemokine Receptor CC-CKR-5." *Nature* 381: 667–73.

Dupré, John (1993). *The Disorder of Things: Metaphysical Foundations of the Disunity of Science.* Cambridge, MA: Harvard University Press.

Earman, John (ed.) (1992). *Inference, Explanation, and Other Philosophical Frustrations*. Berkeley: University of California Press.

Earman, John, and John Roberts (1999). "Ceteris Paribus, There Is No Problem of Provisos." *Synthese* 118: 439–78.

Earman, John, John Roberts, and Sheldon Smith (2002). "Ceteris Paribus Lost." *Erkenntnis* 57: 281–301.

Eells, Ellery (1986). "Probabilistic Causal Interaction." *Philosophy of Science* 53: 52–64.

——— (1987). "Probabilistic Causality: A Reply to John Dupré." *Philosophy of Science* 54: 105–14.

——— (1991). *Probabilistic Causality*. Cambridge: Cambridge University Press.

Eells, Ellery, and Elliott Sober (1983). "Probabilistic Causality and the Question of Transitivity." *Philosophy of Science* 50: 35–57.

Elliott, Kevin (2004). "Error as a Means to Discovery." *Philosophy of Science* 71: 174–97.

Elster, Jon (1983). *Explaining Technological Change: A Case Study in the Philosophy of Science*. Cambridge: Cambridge University Press.

——— (1989). *Nuts and Bolts for the Social Sciences*. Cambridge: Cambridge University Press.

——— (1998). "A Plea for Mechanisms." In Hedström and Swedberg (eds.), pp. 45–73.

Emlen, Douglas, John Hunt, and Leigh Simmons (2005). "The Evolution of Sexual Dimorphism and Male Dimorphism in the Evolution of Beetle Horns: Phylogenetic Evidence for Modularity, Evolutionary Lability, and Constraint." *American Naturalist* 166 (Sciences Module): S42–68.

Eugen-Olsen, Jesper, Astrid Iversen, Peter Garred, Uffe Koppelhus, Court Pedersen, Thomas Benfield, Anne Sorensen, Theresa Katzenstein, Ebbe Dickmeiss, Jan Gerstoft, Peter Skinhoj, Arne Svejgaard, Jens Nielsen, and Bo Hofmann (1997). "Heterozygosity for a Deletion in the CKR5 Gene Leads to Prolonged AIDS-free Survival and Slower CD4 T-cell Decline." *AIDS* 11: 305–10.

Fair, David (1979), "Causation and the Flow of Energy." *Erkenntnis* 14: 219–50.

Fan, Hung, Ross Connor, and Luis Villarreal (2000). *The Biology of AIDS*. Boston: Jones and Bartlett.

Fay, Brian (1983). "General Laws and Explaining Human Behavior." In Daniel Sabia and Jerald Wallulis (eds.), *Changing Social Science*. Albany: SUNY Press, pp. 103–28.

Ferguson, R. Brian (1984). "A Reexamination of the Causes of Northwest Coast Warfare." In Ferguson (ed.), *Warfare, Culture, and Environment*. Orlando, FL: Academic Press, pp. 267–328.

——— (1990). "Blood of the Leviathan: Western Contact and Warfare in Amazonia." *American Ethnologist* 17: 237–57.

——— (1995). *Yanomami Warfare: A Political History*. Santa Fe, NM: School of American Research Press.

Fodor, Jerry (1975). *The Language of Thought*. Cambridge, MA: Harvard University Press.

——— (1991). "You Can Fool Some of the People All of the Time, Everthing Else Being Equal: Hedged Laws and Psychological Explanations." *Mind* 100: 19–34.

——— (1997). "Special Sciences: Still Autonomous After All these Years." *Philosophical Perspectives* 11: 149–64.

Fraker, Thomas, and Rebecca Maynard (1987). "The Adequacy of Comparison Group Designs for Evaluations of Employment-Related Programs." *Journal of Human Resources* 22: 194–227.
Fraser, Hunter (2005). "Modularity and Evolutionary Constraint on Proteins." *Nature Genetics* 37 (4): 351–52.
Friedlander, Daniel, and Gary Burtless (1995). *Five Years After: The Long-Term Effects of Welfare-to-Work Programs*. New York: Russell Sage Foundation.
Friedlander, Daniel, and Philip Robins (1995). "Evaluating Program Evaluation: New Evidence on Commonly Used Nonexperimental Methods." *American Economic Review* 85: 923–37.
Friedman, William, and Joseph Williams (2003). "Modularity of the Angiosperm Female Gametophyte and Its Bearing on the Early Evolution of Endosperm in Flowering Plants." *Evolution* 57: 216–30.
Gambetta, Diego (1998). "Concatenations of Mechanisms." In Hedström and Swedberg (eds.).
Garfinkel, Irwin, Charles Manski, and Charles Michalopoulos (1992). "Micro Experiments and Macro Effects." In Manski and Garfinkel (eds.), pp. 253–76.
Gartner, S., P. Markovits, D. Markovitz, M. Kaplan, R. Gallo, and M. Popovic (1986). "The Role of Mononuclear Phagocytes in HTLV-III/LAV Infection." *Science* 233: 215–19.
Geiger, Dan, and David Heckerman (1991). "Advances in Probabilistic Reasoning." In *Uncertainty in Artificial Intelligence: Proceedings of the Seventh Conference*. San Mateo, CA: Morgan Kaufman, pp. 118–26.
George, Alexander, and Andrew Bennett (2005). *Case Studies and Theory Development in the Social Sciences*. Cambridge, MA: MIT Press.
Gigerentzer, Gerd (2000). *Adaptive Thinking: Rationality in the Real World*. Oxford: Oxford University Press.
Glennan, Stuart (1996). "Mechanisms and the Nature of Causation." *Erkenntnis* 44: 49–71.
——— (1997). "Capacities, Universality, and Singularity." *Philosophy of Science* 64: 605–26.
——— (2002). "Rethinking Mechanistic Explanation." *Philosophy of Science* 69: S342–53.
——— (2005). "Modeling Mechanisms." *Studies in History and Philosophy of Biological and Biomedical Science* 36: 443–64.
Glushakova, Svetlana, Jean-Charles Grivel, Wendy Fitzgerald, Andrew Sylwester, Joshua Zimmerberg, and Leonid Margolis (1998). "Evidence for the HIV-1 Phenotype Switch as a Causal Factor in Acquired Immunodeficiency." *Nature Medicine* 4: 346–49.
Glymour, Clark (1997). "A Review of Recent Work on the Foundations of Causal Inference." In V. R. McKim and S. P. Turner (eds.), pp. 201–48.
——— (1999). "Rabbit Hunting." *Synthese* 121: 55–78.
——— (2001). *The Mind's Arrows: Bayes Nets and Graphical Causal Models in Psychology*. Cambridge, MA: MIT Press.
——— (2002). "A Semantics and Methodology for Ceteris Paribus Hypotheses." *Erkenntnis* 57: 395–405.
Glymour, Clark, and Gregory Cooper (eds.) (1999). *Computation, Causation, and Discovery*. Cambridge, MA: MIT Press.
Gold, Lois, Neela Manley, and Bruce Ames (1992). "Extrapolation of Cacinogenicity Between Species: Qualitative and Quantitative Factors." *Risk Analysis* 12: 579–88.

Goodman, Nelson (1954). *Fact, Fiction and Forecast.* Cambridge, MA: Harvard University Press.

Grether, David, and Charles Plott (1979). "Economic Theory of Choice and the Preference Reversal Phenomenon." *American Economic Review* 69: 623–38.

Guala, Francesco (2005). *The Methodology of Experimental Economics.* Cambridge: Cambridge University Press.

Gueron, Judith, and Edward Pauly (1991). *From Welfare to Work.* New York: Russell Sage Foundation.

Halpern, John, and Judea Pearl (2005). "Causes and Explanations: A Structural-Model Approach. Part I: Causes." *British Journal for the Philosophy of Science* 56: 843–87.

Hannan, Michael, and John Freeman (1989). *Organizational Ecology.* Cambridge, MA: Harvard University Press.

Hanson, Norwood (1958). *Patterns of Discovery.* Cambridge: Cambridge University Press.

Harris, Marvin (1977). *Cannibals and Kings: The Origins of Cultures.* New York: Random House.

——— (1984). "Animal Capture and Yanomamo Warfare: Retrospect and New Evidence." *Journal of Anthropological Research* 40: 183–201.

——— (1996). "Yanomami Warfare: A Political History—Ferguson, R. B." *Human Ecology* 24: 413–16.

Hausman, Daniel (1992). *The Inexact and Separate Science of Economics.* Cambridge: Cambridge University Press.

——— (1998). *Causal Asymmetries.* Cambridge: Cambridge University Press.

Hausman, Daniel, and James Woodward (1999). "Independence, Invariance and the Causal Markov Condition." *British Journal for the Philosophy of Science* 50: 521–83.

——— (2004). "Modularity and the Causal Markov Condition: A Restatement." *British Journal for the Philosophy of Science* 55: 147–61.

Heckman, James (1992). "Randomization and Social Policy Evaluation." In Manski and Garfinkel (eds.), pp. 201–30.

——— (1996). "Identification of Causal Effects Using Instrumental Variables: Comment." *Journal of the American Statistical Association* 91 (434): 459–62.

Heckman, James, and V. Joseph Hotz (1989). "Choosing Among Alternative Nonexperimental Methods for Estimating the Impact of Social Programs: The Case of Manpower Training." *Journal of the American Statistical Association* 84: 862–74.

Hedström, Peter, and Richard Swedberg (eds.) (1998). *Social Mechanisms: An Analytical Approach to Social Theory.* Cambridge: Cambridge University Press.

Heinen, H. D., and B. Illius (1996). "The Last Days of El Dorado: A Review Essay on Yanomami Warfare." *Anthropos* 91: 552–60.

Hempel, Carl (1965). *Aspects of Scientific Explanation, and Other Essays in the Philosophy of Science.* New York: Free Press.

Hengstler, Jan, M. S. Bogdanffy, H. M. Bolt, and F. Oesch (2003). "Challenging Dogma: Thresholds for Genotoxic Carcinogens? The Case of Vinyl Acetate." *Annual Review of Pharmacological Toxicology* 43: 485–520.

Hengstler, Jan, Bart van der Burg, Pablo Steinberg, and Franz Oesch (1999). "Interspecies Differences in Cancer Susceptibility and Toxicity." *Drug Metabolism Reviews* 31: 917–70.

Hernes, Gudmund (1998). "Real Virtuality." In Hedström and Swedberg (eds.), pp. 74–101.Hesslow, Germund (1976). "Two Notes on the Probabilistic Approach to Causality." *Philosophy of Science* 43: 290–92.

Hitchcock, Christopher (1993). "A Generalized Theory of Causal Relevance." *Synthese* 97: 335–64.

—— (1995). "The Mishap at Reichenbach Fall: Singular vs. General Causation." *Philosophical Studies* 78: 257–91.

—— (2003). "Of Humean Bondage." *British Journal for the Philosophy of Science* 54: 1–25.

Holland, Paul (1986). "Statistics and Causal Influence." *Journal of the American Statistical Association* 81 (396): 946–60.

Holt, Charles (1986). "Preference Reversals and the Independence Axiom." *American Economic Review* 76: 508–15.

Hooker, C. A. (2004). "Asymptotics, Reduction and Emergence." *British Journal for the Philosophy of Science* 55: 435–80.

Hoover, Kevin (2001). *Causality in Macroeconomics*. Cambridge: Cambridge University Press.

—— (2003). "Nonstationary Time Series, Cointegration, and the Principle of the Common Cause." *British Journal for the Philosophy of Science* 54: 527–51.

Hotz, V. Joseph (1992). "Designing an Evaluation of the Job Training Partnership Act." In Manski and Garfinkel (eds.).

Hoynes, Hilary (2000). "The Employment, Earnings, and Income of Less Skilled Workers over the Business Cycle." In Card and Blank (eds.).

Hull, David (1972). "Reduction in Genetics—Biology or Philosophy?" *Philosophy of Science* 38: 491–99.

—— (1974). *Philosophy of Biological Science*. Englewood Cliffs, NJ: Prentice-Hall.

Humphreys, Paul (1989). *The Chances of Explanation*. Princeton, NJ: Princeton University Press.

—— (1997). "How Properties Emerge." *Philosophy of Science* 64: 1–17.

Jones, Todd (1996). "Methodological Individualism in Proper Perspective." *Behavior and Philosophy* 24: 119–28.

—— (1998). "Interpretive Social Science and the 'Native's Point of View': A Closer Look." *Philosophy of the Social Sciences* 28: 32–68.

—— (1999). "FIC Descriptions and Interpretive Social Science: Should Philosophers Roll Their Eyes?" *Journal for the Theory of Social Behavior* 29: 337–69.

Joyner, Alexandra (2000). *Gene Targeting: A Practical Approach*. Oxford: Oxford University Press.

Kagel, John, and Alvin Roths (eds.). *The Handbook of Experimental Economics*. Princeton, NJ: Princeton University Press.

Kahneman, Daniel, Paul Slovic, and Amos Tversky (eds.) (1982). *Judgment Under Uncertainty: Heuristics and Biases*. Cambridge: Cambridge University Press.

Kalichman, Seth (1998). *Understanding AIDS*. 2nd ed. Washington, DC: American Psychological Association.

Kashtan, Nadav, and Uri Alon (2005). "Spontaneous Evolution of Modularity and Network Motifs." *Proceedings of the National Academy of Sciences* 102 (39): 13773–78.

Kelly, Kevin (1996). *The Logic of Reliable Inquiry*. Oxford: Oxford University Press.

Kew, Michael (2003). "Synergistic Interaction Between Aflatoxin B_1 and Hepatitis B Virus in Hepatocarcinogen." *Liver International* 23: 405–9.

Kim, Jaegwon (1992). "Multiple Realization and the Metaphysics of Reduction." *Philosophy and Phenomenological Research* 52: 1–26.

——— (1999). "Making Sense of Emergence." *Philosophical Studies* 95: 3–36.

Kincaid, Harold (1990). "Molecular Biology and the Unity of Science." *Philosophy of Science* 57: 575–93.

——— (1996). *Philosophical Foundations of the Social Sciences*. Cambridge: Cambridge University Press.

——— (1997). *Individualism and the Unity of Science: Essays on Reduction, Explanation, and the Special Sciences*. Lanham, MD: Rowman & Littlefield.

Kitcher, Philip (1984). "1953 and All That, a Tale of Two Sciences." *Philosophical Review* 93: 335–73.

——— (1999). "The Hegemony of Molecular Biology." *Biology and Philosophy* 14: 196–210.

——— (2001). *Science, Truth, and Democracy*. Oxford: Oxford University Press.

——— (2002). "Reply to Helen Longino." *Philosophy of Science* 69: 569–72.

Koster, J. T. A. (1996). "Markov Properties of Non-Recursive Models." *Annals of Statistics* 24: 2148–78.

Kuhn, Thomas (1977). *The Essential Tension: Selected Studies in Scientific Tradition and Change*. Chicago: University of Chicago Press.

Kurihara, Laurie Jo, Tateki Kikuchi, Keiji Wada, and Shirley M. Tilghman (2001). "Loss of Uch-L1 and Uch-L3 Leads to Neurodegeneration, Posterior Paralysis and Dysphagia." *Human Molecular Genetics* 10: 1963–70.

Kushnir, Tamar, and Alison Gopnik (2005). "Young Children Infer Causal Strength from Probabilities and Interventions." *Psychological Science* 16: 678–83.

Kvasnicka, Vladimir, and Jiri Pospichal (2002). "Emergence of Modularity in Genotype-Phenotype Mappings." *Artificial Life* 8: 295–310.

LaFollette, Hugh, and Niall Shanks (1993a). "Animal Models in Biomedical Research: Some Epistemological Worries." *Public Affairs Quarterly* 7: 113–30.

——— (1993b). "The Intact Systems Argument: Problems with the Standard Defense of Animal Extrapolation." *Southern Journal of Philosophy* 31: 323–33.

——— (1995). "Two Models of Models in Biomedical Research." *Philosophical Quarterly* 45: 141–60.

——— (1996). *Brute Science: Dilemmas of Animal Experimentation*. New York: Routledge.

LaLonde, Robert (1986). "Evaluating the Econometric Evaluations of Training Programs with Experimental Data." *American Economic Review* 4: 604–20.

Lange, Marc (1993). "Natural Laws and the Problem of Provisos." *Erkenntnis* 38: 233–48.

——— (2000). *Natural Laws in Scientific Practice*. Oxford: Oxford University Press.

——— (2002). "Who's Afraid of Ceteris Paribus Laws? Or: How I Learned to Stop Worrying and Love Them." *Erkenntnis* 57: 407–23.

Langlois, Richard (2002). "Modularity in Technology and Organization." *Journal of Economic Behavior and Organization* 49: 19–37.

Lewis, David (1970). "How to Define Theoretical Terms." *Journal of Philosophy* 67: 427–46.

——— (1994). "Reduction and Mind." In Samuel Guttenplan (ed.), *A Companion to the Philosophy of Mind*. Oxford: Blackwell, pp. 412–31.

Liljegren, Sarah, Gary Ditta, Yuval Eshed, Beth Savidge, John Bowman, and Martin Yanofsky (2000). "Shatterproof MADS-box Genes Control Seed Dispersal in *Arabidopsis*." *Nature* 404: 766–70.

Lipson, Hod, Jordan Pollack, and Nam Suh (2002). "On the Origin of Modular Variation," *Evolution* 56: 1549–56.
Lipton, Peter (1999). "All Else Being Equal." *Philosophy* 74: 155–68.
Little, Daniel (1991). *Varieties of Social Explanation: An Introduction to the Philosophy of Social Science*. Boulder, CO: Westview Press.
——— (1995a). "Causal Explanation in the Social Sciences." *Southern Journal of Philosophy* 34 (Supplement): 31–56.
——— (ed.) (1995b). *On the Reliability of Economic Models*. Boston: Kluwer Academic.
——— (1998). *Microfoundations, Method, and Causation*. New Brunswick, NJ: Transaction Publishers.
Liu, Rong, William Paxton, Sunny Choe, Daniel Ceradini, Scott Martin, Richard Horuk, Marcy MacDonald, Heidi Stuhlmann, Richard Koup, and Nathaniel Landau (1996). "Homozygous Defect in HIV-1 Coreceptor Accounts for Resistance of Some Multiply-Exposed Individuals to HIV-1 Infection." *Cell* 86: 367–77.
Longino, Helen (1990). *Science as Social Knowledge: Values and Objectivity in Scientific Inquiry*. Princeton, NJ: Princeton University Press.
——— (2000). "Toward an Epistemology for Biological Pluralism." In R. Creath and J. Maienschein (eds.), *Biology and Epistemology*. Cambridge: Cambridge University Press.
——— (2002a). *The Fate of Knowledge*. Princeton, NJ: Princeton University Press.
——— (2002b). "Reply to Philip Kitcher." *Philosophy of Science* 69: 573–77.
Lucas, Robert (1981). *Studies in Business Cycle Theory*. Cambridge, MA: MIT Press.
Mabee, Paula, Patricia Crotwell, Nathan Bird, and Ann Burke (2002). "Evolution of Median Fin Modules in the Axial Skeleton of Fishes." *Journal of Experimental Zoology* 294: 77–90.
Machamer, Peter, Lindley Darden, and Carl Craver (2000). "Thinking About Mechanisms." *Philosophy of Science* 67: 1–25.
Machamer, Peter, Rick Grush, and Peter McLaughlin (eds.) (2001). *Theory and Method in the Neurosciences*. Pittsburgh, PA: University of Pittsburgh Press.
Maddon, Paul, Angus Dalgleish, Stephen McDougal, Paul Clapham, Robin Weiss, and Richard Axel (1986). "The T4 Gene Encodes the AIDS Virus Receptor and Is Expressed in the Immune System and the Brain." *Cell* 47: 333–48.
Malinowski, Bronislaw (1935). *Coral Gardens and Their Magic*. New York: American Book Co.
Manski, Charles, and Irwin Garfinkel (eds.) (1992). *Evaluating Welfare and Training Programs*. Cambridge, MA: Harvard University Press.
Martin, Michael (1993). "Geertz and the Interpretive Approach to Anthropology." *Synthese* 97: 269–86.
Mayntz, Renate (2004). "Mechanisms in the Analysis of Social Macro-Phenomena." *Philosophy of the Social Sciences* 34: 237–59.
McGlynn, Katherine, Kent Hunter, Thomas LeVoyer, Jessica Roush, Philip Wise, Rita Michielli, Fu-Min Shen, Alison Evans, W. Thomas London, and Kenneth Buetow (2003). "Susceptibility to Aflatoxin B_1-related Primary Hepatocellular Carcinoma in Mice and Humans." *Cancer Research* 63: 4594–601.
McKim, Vaughn, and Stephen Turner (eds.) (1997). *Causality in Crisis? Statistical Methods and the Search for Causal Knowledge in the Social Sciences*. Notre Dame, IN: University of Notre Dame Press.
Meek, Christopher (1995). "Strong Completeness and Faithfulness in Bayesian Networks." In P. Besnard (ed.), *Uncertainty in Artificial Intelligence: Proceedings of the Eleventh Conference*. San Francisco: Morgan Kaufman, pp. 411–18.

Mellor, Hugh (1988). "Raising the Chances of Effects." In James H. Fetzer (ed.), *Probability and Causality: Essays in Honor of Wesley C. Salmon*. Boston: D. Reidel, pp. 229–39.

Menzies, Peter (1996). "Probabilistic Causation and the Pre-emption Problem." *Mind* 105: 85–117.

Menzies, Peter, and Huw Price (1993). "Causation as a Secondary Quality." *British Journal for the Philosophy of Science* 44: 187–203.

Michalopoulos, Charles, Howard Bloom, and Carolyn Hill (2004). "Can Propensity-Score Methods Match the Findings from a Random Assignment Evaluation of Mandatory Welfare-to-Work Programs?" *Review of Economics and Statistics* 86: 156–79.

Mitchell, Sandra (1997). "Pragmatic Laws." *Philosophy of Science* 64 (Supplement): S468–79.

——— (2000). "Dimensions of Scientific Law." *Philosophy of Science* 67: 242–65.

——— (2002a). "Ceteris Paribus—An Inadequate Representation for Biological Contingency." *Erkenntnis* 329–50.

——— (2002b). "Integrative Pluralism." *Biology and Philosophy* 17: 55–70.

——— (2003). *Biological Complexity and Integrative Pluralism*. Cambridge: Cambridge University Press.

Morreau, Michael (1999). "Other Things Being Equal." *Philosophical Studies* 96: 164–82.

Morrison, Margaret (1995). "Capacities, Tendencies and the Problem of Singular Causes." *Philosophy and Phenomenological Research* 55: 163–68.

Mott, Peter (1992). "Fodor and Ceteris Paribus Laws." *Mind* 101: 335–46.

Nagel, Ernest (1961). *The Structure of Science: Problems in the Logic of Scientific Discovery*. London: Routledge and Kegan Paul.

——— (1979). *Teleology Revisited and Other Essays in the Philosophy and History of Science*. New York: Columbia University Press.

Neopolitan, Richard (2004). *Learning Bayesian Networks*. Upper Saddle River, NJ: Pearson Prentice-Hall.

Oppenheim, Paul, and Hilary Putnam (1958). "The Unity of Science as a Working Hypothesis." In Herbert Feigl (ed.), *Minnesota Studies in the Philosophy of Science*, vol. 2. Minneapolis: University of Minnesota Press.

Ozawa, Martha, and Hong-Sik Yoon (2005). " 'Leavers' from TANF to AFDC: How Do They Fare Economically?" *Social Work* 50: 239–49.

Paige, Jeffrey (1975). *Agrarian Revolution: Social Movements and Export Agriculture in the Underdeveloped World*. New York: Free Press.

Papineau, David (1978). *For Science in the Social Sciences*. London: Macmillan.

——— (1996). "Theory-Dependent Terms." *Philosophy of Science* 63: 1–20.

Pearl, Judea (1988). *Probabilistic Reasoning in Intelligent Systems*. San Mateo, CA: Morgan Kaufman.

——— (1998). "TETRAD and SEM." *Multivariate Behavioral Research* 33: 119–28.

——— (2000). *Causality: Models, Reasoning, and Inference*. Cambridge: Cambridge University Press.

Pearson, Helen (2002). "Surviving a Knockout Blow." *Nature* 415: 8–9.

Persky, Joseph (1990). "Retrospectives: Ceteris Paribus." *Journal of Economic Perspectives* 4: 187–93.

Pietroski, Paul, and George Rey (1995). "When Other Things Aren't Equal: Saving Ceteris Paribus Laws from Vacuity." *British Journal for the Philosophy of Science* 46: 81–110.

Pollock, Donald (1997). "Yanomami Warfare: A Political History—Ferguson, R. B." *Ethnohistory* 44: 191–93.

Popper, Karl ([1959] 2000). *The Logic of Scientific Discovery*. New York: Routledge.

——— ([1963] 1989). *Conjectures and Refutations*. 5th ed., rev. New York: Routledge.

Putnam, Hilary (1975). *Mind, Language and Reality: Philosophical Papers*, vol. 2. New York: Cambridge University Press.

Radetsky, Peter (1997). "Immune to a Plague: Gene for Immunity to AIDS Discovered." *Discover* 18 (6): 60–67.

Ramsey, Frank P. (1954). "Theories." In R. B. Braithwaite (ed.).

Reedy-Maschner, Katherine, and Herbert Maschner (1999). "Marauding Middlemen: Western Expansion and Violent Conflict in the Subarctic." *Ethnohistory* 46: 704–43.

Reichenbach, Hans (1938). *Experience and Prediction*. Chicago: University of Chicago Press.

——— (1956). *The Direction of Time*, edited by Maria Reichenbach. Berkeley: University of California Press.

Richardson, Thomas, and Peter Spirtes (1999). "Automated Discovery of Linear Feedback Models." In Glymour and Cooper (eds.).

Richerson, Peter, and Robert Boyd (2005). *Not by Genes Alone: How Culture Transformed Human Evolution*. Chicago: University of Chicago Press.

Rivière, Peter (1996). "Yanomami Warfare: A Political History—Ferguson, R. B." *Journal of Latin American Studies* 28: 262–63.

Robbins, Suzanne, and Holly Barcus (2004). "Welfare Reform and Economic and Housing Capacities for Low-Income Households, 1997–1999." *Policy Studies Journal* 32: 439–60.

Robinson, Joseph (1992). "Aims and Achievements of the Reductionist Approach in Biochemistry/Molecular Biology/Cell Biology: A Response to Kincaid." *Philosophy of Science* 59: 465–70.

Rosenbaum, Paul (2002). *Observational Studies*, 2nd ed. New York: Springer-Verlag.

Rosenberg, Alexander (1985). *The Structure of Biological Science*. New York: Cambridge University Press.

——— (1997). "Reductionism Redux: Computing the Embryo." *Biology and Philosophy* 68: 445–70.

——— (2001). "Reductionism in a Historical Science." *Philosophy of Science* 68: 135–63.

Roth, Alvin (1995). "Introduction to Experimental Economics." In John Kagel and Alvin Roth (eds.), *The Handbook of Experimental Economics*. Princeton, NJ: Princeton University Press.

Ruphy, Stéphanie (2003). "Is the World Really 'Dappled'?: A Response to Cartwright's Charge Against 'Cross-Wise Reduction.'" *Philosophy of Science* 70: 57–67.

Salmon, Wesley (1984). *Scientific Explanation and the Causal Structure of the World*. Princeton, NJ: Princeton University Press.

Samson, Michel, Frédéric Libert, Benjamin Doranz, Joseph Rucker, Corinne Liesnard, Claire-Michèle Farber, Sentob Saragosti, Claudine Lapouméroulie, Jacqueline Cognaux, Christine Forceille, Gaetan Muyldermans, Chris Verhofstede, Guy Burtonboy, Michel Georges, Tsuneo Imai, Shalini Rana, Yanji Yi, Robert J. Smyth, Ronald G. Collman, Robert Doms, Gilbert Vassart, and Marc Parmentier (1996). "Resistance to HIV-1 Infection in Caucasian Individuals Bearing Mutant Alleles of the CCR-5 Chemokine Receptor Gene." *Nature* 382: 722–25.

Sarkar, Sahotra (1998). *Genetics and Reductionism*. Cambridge: Cambridge University Press.

Scarff, Katrina, Kheng Ung, Harshal Nandurkar, Peter Crack, Catherina Bird, and Phillip Bird (2004). "Targeted Disruption of SPI3/Serpinb6 Does Not Result in Developmental or Growth Defects, Leukocyte Dysfunction, or Susceptibility to Stroke." *Molecular and Cellular Biology* 24: 4075–82.

Schaffner, Kenneth (1967). "Approaches to Reduction." *Philosophy of Science* 34: 137–47.

——— (1969). "The Watson-Crick Model and Reductionism." *British Journal for the Philosophy of Science* 20: 325–48.

——— (1993a). "Theory Structure, Reduction, and Disciplinary Integration in Biology." *Biology and Philosophy* 8: 319–47.

——— (1993b). *Discovery and Explanation in Biology and Medicine*. Chicago: University of Chicago Press.

——— (2001). "Extrapolation from Animal Models: Social Life, Sex, and Super Models." In Machamer, Grush, and McLaughlin (eds.).

Scheines, Richard (1997). "An Introduction to Causal Inference." In McKim and Turner (eds.), pp. 185–99.

Scheines, Richard, Peter Spirtes, Clark Glymour, Cristopher Meek, and Thomas Richardson (1998). "Reply to Comments." *Multivariate Behavioral Research* 33: 165–80.

Schelling, Thomas (1978). *Micromotives and Macrobehavior*. New York: Norton.

——— (1998). "Social Mechanisms and Social Dynamics." In Hedström and Swedberg (eds.), pp. 32–44.

Schiffer, Stephen (1991). "Ceteris Paribus Laws." *Mind* 100: 1–17.

Schlosser, Gerhard, and Günter Wagner (2004). "Introduction: The Modularity Concept in Developmental and Evolutionary Biology." In Schlosser and Wagner (eds.).

Schlosser, Gerhard, and Günter Wagner (eds.) (2004). *Modularity in Development and Evolution*. Chicago: University of Chicago Press.

Schurz, Gerhard (2001a). "What Is 'Normal'? An Evolution-Theoretic Foundation for Normic Laws and Their Relation to Statistical Normality." *Philosophy of Science* 68: 476–97.

Schurz, Gerhard (2001b). "Pietroski and Rey on Ceteris Paribus Laws." *British Journal for the Philosophy of Science* 52: 359–70.

——— (2002). "Ceteris Paribus Laws: Classification and Deconstruction." *Erkenntnis* 57: 351–72.

Scott, James (1985). *Weapons of the Weak: Everyday Forms of Peasant Resistance*. New Haven, CT: Yale University Press.

Searle, John (1984). *Minds, Brains and Science*. Cambridge, MA: Harvard University Press.

Segal, Uzi (1988). "Does the Preference Reversal Phenomenon Necessarily Contradict the Independence Axiom?" *American Economic Review* 78: 233–36.

Seidl, Christian (2002). "Preference Reversal." *Journal of Economic Surveys* 16: 621–55.

Selten, Reinhard (2001). "What Is Bounded Rationality?" In G. Gigerenzer and R. Selten (eds.), *Bounded Rationality: The Adaptive Toolbox*. Cambridge, MA: MIT Press, pp. 13–36.

Shipley, Bill (2000). *Cause and Correlation in Biology: A User's Guide to Path Analysis, Structural Equations and Causal Inference*. Cambridge: Cambridge University Press.

Simon, Herbert (1962). "The Architecture of Complexity." *Proceedings of the American Philosophical Society* 106 (6): 467–82.

——— (1998). "Does Discovery Have a Logic?" In H. Keuth (ed.), *Karl Popper, Logik der Forschung*. Berlin: Akademie Verlag, pp. 235–48.

Skyrms, Brian (1980). *Causal Necessity: A Pragmatic Investigation of the Necessity of Laws*. New Haven, CT: Yale University Press.

Slovic, Paul (1995). "The Construction of Preference." *American Psychologist* 50: 364–71.

Smith, Sheldon (2002). "Violated Laws, Ceteris Paribus Clauses, and Capacities." *Synthese* 130: 235–64.

Sober, Elliott (1999). "The Multiple Realizability Argument Against Reductionism." *Philosophy of Science* 66: 542–64.

——— (2001). "Venetian Sea Levels, British Bread Prices, and the Principle of the Common Cause." *British Journal for the Philosophy of Science* 52: 331–46.

Sohn, Soojung, Iris Jaitovitch-Groisman, Naciba Benlimame, Jacques Galipeau, Gerald Batist, and Moulay Alaoui-Jamali (2000). "Retroviral Expression of the Hepatitis B Virus x Gene Promotes Liver Cell Susceptibility to Carcinogen-Induced Site Specific Mutagenesis." *Mutation Research* 460: 17–28.

Spirtes, Peter (1995). "Directed Cyclic Graphical Representation of Feedback Models." In *Uncertainty in Artificial Intelligence: Proceedings of the Eleventh Conference*. San Francisco: Morgan Kaufman, pp. 491–98.

Spirtes, Peter, Clark Glymour, and Richard Scheines (1993). *Causation, Prediction, and Search*. New York: Springer-Verlag.

——— (2000). *Causation, Prediction, and Search*, 2nd ed. Cambridge, MA: MIT Press.

Spohn, Wolfgang (2002). "Laws, Ceteris Paribus Conditions, and the Dynamics of Belief." *Erkenntnis* 57: 373–94.

Sponsel, Leslie (1998). "Yanomami: An Arena of Conflict and Aggression in the Amazon." *Aggressive Behavior* 24: 97–122.

Steel, Daniel (1999). "Trade Goods and Jívaro Warfare: The Shuar 1850–1957, and the Achuar 1940–1978." *Ethnohistory* 46: 745–76.

——— (2003). "Making Time Stand Still: A Response to Sober's Counter-example to the Principle of the Common Cause." *British Journal for the Philosophy of Science* 54: 309–17.

——— (2004). "Can a Reductionist Be a Pluralist." *Biology and Philosophy* 16: 55–73.

——— (2005). "Indeterminism and the Causal Markov Condition." *British Journal for the Philosophy of Science* 56: 3–26.

——— (2006). "Homogeneity, Selection, and the Faithfulness Condition." *Minds and Machines* 16: 303–17.

Sterelny, Kim, and Paul Griffiths (1999). *Sex and Death: An Introduction to Philosophy of Biology*. Chicago: University of Chicago Press.

Stinchcombe, Arthur (1991). "The Conditions of Fruitfulness of Theorizing About Mechanisms in Social Science." *Philosophy of the Social Sciences* 21: 367–88.

Stine, Gerald (2000). *AIDS Update 2000*. Upper Saddle River, NJ: Prentice-Hall.

Stirzaker, David (2003). *Elementary Probability*, 2nd ed. Cambridge: Cambridge University Press.

Stouffer, Samuel (1949). *The American Soldier*. New York: Wiley.

Suppe, Frederick (ed.) (1974). *The Structure of Scientific Theories*. Chicago: University of Illinois Press.

Suppes, Patrick (1970). *A Probabilistic Theory of Causality.* Amsterdam: North-Holland.

Tabery, James (2004). "Synthesizing Activities and Interactions in the Concept of a Mechanism." *Philosophy of Science* 71: 1–15.

Taylor, Charles (1971). "Interpretation and the Sciences of Man." *Review of Metaphysics* 25: 3–51.

Tooley, Michael (1987). *Causation: A Realist Approach.* Oxford: Clarendon Press.

Tversky, Amos, Paul Slovic, and Daniel Kahneman (1990). "The Causes of Preference Reversal." *American Economic Review* 80: 204–17.

Tymms, Martin, and Ismail Kola (eds.) (2001). *Gene Knockout Protocols.* Totowa, NJ: Humana Press.

Venn, John (1962). *The Logic of Chance*, 4th ed. New York: Chelsea Publishing Co.

Wagner, Günter, and Lee Altenberg (1996). "Complex Adaptations and the Evolution of Evolvability." *Evolution* 50: 967–76.

Watanabe, Karen, Frédéric Dois, and Lauren Zeise (1992). "Interspecies Extrapolation: A Reexamination of Acute Toxicity Data." *Risk Analysis* 12: 301–10.

Waters, Kenneth (1990). "Why the Anti-Reductionist Consensus Won't Survive: The Case of Classical Mendelian Genetics." In A. Fine, M. Forbes, and L. Wessels (eds.), *PSA 1990: Proceedings of the 1990 Biennial Meeting of the Philosophy of Science Association*, vol. 1. East Lansing, MI: Philosophy of Science Association.

Watson, Richard, and Jordan Pollack (2005). "Modular Independency in Complex Dynamical Systems." *Artificial Life* 11: 445–57.

Weber, Marcel (2001). "Under the Lamppost: Commentary on Schaffner." In Machamer, Grush, and McLaughlin (eds.), pp. 213–49.

——— (2005). *Philosophy of Experimental Biology.* Cambridge: Cambridge University Press.

Weitzenfeld, Julian (1984). "Valid Reasoning by Analogy." *Philosophy of Science* 51: 137–49.

Wimsatt, William (1976). "Reductive Explanation: A Functionalist Account." In R. S. Cohen et al. (eds.), *PSA 1974.* Dordrecht: D. Reidel., pp. 671–710.

——— (1979). "Reduction and Reductionism." In P. D. Asquith and H. Kyburg (eds.), *Current Research in the Philosophy of Science.* East Lansing, MI: Philosophy of Science Association, pp. 352–77.

Wimsatt, William (1998). "Simple Systems and Phylogenetic Diversity." *Philosophy of Science* 65: 267–75.

Winship, Christopher, and Stephen Morgan (1999). "The Estimation of Causal Effects from Observational Data." *Annual Review of Sociology* 25: 659–707.

Wogan, Gerald (1992). "Aflatoxin Carcinogenesis: Interspecies Potency Differences and Relevance for Human Risk Assessment." In R. D'Amato, T. Slaga, W. Farland, and C. Henry (eds.), *Relevance of Animal Studies to the Evaluation of Human Cancer Risk.* New York: Wiley, pp. 123–38.

Wolfe, Barbara (2002). "Incentives, Challenges, and Dilemmas of TANF: A Case Study." *Journal of Policy Analysis and Management* 21: 577–86.

Woodward, James (1998). "Causal Independence and Faithfulness." *Multivariate Behavioral Research* 33: 129–48.

——— (1999). "Causal Interpretation in Systems of Equations." *Synthese* 121: 199–257.

——— (2000). "Explanation and Invariance in the Special Sciences." *British Journal of the Philosophy of Science* 51: 197–254.

Woodward, James(2001). "Law and Explanation in Biology: Invariance Is the Kind of Stability That Matters." *Philosophy of Science* 68: 1–20.

——— (2002a). "What Is a Mechanism? A Counterfactual Account." *Philosophy of Science* 69 (Supplement): 366–78.

——— (2002b). "There Is No Such Thing as a Ceteris Paribus Law." *Erkentnnis* 57: 303–28.

——— (2003). *Making Things Happen: A Causal Theory of Explanation*. Oxford: Oxford University Press.

Wu, Xifeng, Jun Gu, Yehuda Patt, Manal Hassan, Margaret Spitz, R. Palmer Beasley, and Lu-Yu Hwang (1998). "Mutagen Sensitivity as a Susceptibility Marker for Human Hepatocellular Carcinoma." *Cancer Epidemiology* 7: 567–70.

Zhu, T., H. Mo, N. Wang, D. Nam, Y. Cao, R. Koup, and D. Ho (1993). "Genotypic and Phenotypic Characterization of HIV-1 in Patients with Primary Infection." *Science* 261: 1179–81.

Index

ACROSS THE BOUNDARIES

ENVIRONMENTAL ETHICS AND SCIENCE POLICY SERIES
General Editor: Kristin Shrader-Frechette

IN NATURE'S INTERESTS?
Interests, Animal Rights, and Environmental Ethics
Gary E. Varner

PRIVATIZING PUBLIC LANDS
Scott Lehman

DEMOCRACY, RISK, AND THE COMMUNITY
Technological Hazards and the Evolution of Liberalism
Richard P. Hiskes

ENVIRONMENTAL JUSTICE
Creating Equality, Reclaiming Democracy
Kristin Shrader-Frechette

ACROSS THE BOUNDARIES
Extrapolation in Biology and Social Science
Daniel P. Steel

ACROSS THE BOUNDARIES

Extrapolation in Biology and Social Science

Daniel P. Steel

OXFORD
UNIVERSITY PRESS
2008

OXFORD
UNIVERSITY PRESS

Oxford University Press, Inc., publishes works that further
Oxford University's objective of excellence
in research, scholarship, and education.

Oxford New York
Auckland Cape Town Dar es Salaam Hong Kong Karachi
Kuala Lumpur Madrid Melbourne Mexico City Nairobi
New Delhi Shanghai Taipei Toronto

With offices in
Argentina Austria Brazil Chile Czech Republic France Greece
Guatemala Hungary Italy Japan Poland Portugal Singapore
South Korea Switzerland Thailand Turkey Ukraine Vietnam

Published by Oxford University Press, Inc.
198 Madison Avenue, New York, New York 10016
www.oup.com

Steel, Daniel, 1970–
Across the boundaries : extrapolation in biology and social science / Daniel Steel.
p. ; cm. — (Environmental ethics and science policy series)
Includes bibliographical references.
ISBN 978-0-19-533144-8
1. Medicine—Research—Philosophy. 2. Biology—Research—Philosophy.
3. Social sciences—Research—Philosophy. 4. Extrapolation. 5. Causality. I. Title.
II. Series: Environmental ethics and science policy.
[DNLM: 1. Causality. 2. Physiology, Comparative. 3. Population Characteristics.
4. Psychology, Comparative. 5. Public Policy. 6. Research Design. QT 4 S813a 2007]
R852.S72 2007
610.1—dc22 2007007050

9 8 7 6 5 4 3 2 1

Printed in the United States of America
on acid-free paper

For Lynne Anne Steel

Preface

The prehistory of this book began in the late 1990s, when I was a graduate student in the Department of History and Philosophy of Science at the University of Pittsburgh in search of a dissertation topic. I wanted to write something about problems relating to learning about cause and effect in social science, but was having trouble finding a topic that hadn't already been done almost to death and that didn't require mastering impossibly intimidating mathematics. By chance, I stumbled across an edited volume in the Hillman Library titled *Evaluating Welfare and Training Programs* (Manski and Garfinkel 1992), which contained insightful and stimulating discussions of methodological difficulties relating to extrapolating results of pilot studies of welfare-to-work programs. One appealing idea that occurred to me, as well as to some contributors to *Evaluating Welfare and Training Programs,* was that reliable extrapolation often relies upon knowledge of mechanisms linking cause and effect and of factors capable of interfering with those mechanisms. But despite, or perhaps because of, its intuitive obviousness, I had difficulty finding any attempt to elaborate, clarify, or otherwise work out the details of this idea. Here at last was a dissertation topic par excellence! I decided to approach the issue by looking at biological examples that I presumed would best exemplify how mechanisms-based extrapolation could proceed. Figuring out how extrapolation worked in biological cases turned out to be a lot more complicated than I had imagined, and ended up comprising the bulk of the dissertation. As the thoroughly revised, rewritten, and rethought descendant of that dissertation, this book retains the central role of biological examples. Looking back on close to a decade of effort on the topic of extrapolation, I have a sense both of satisfaction in progress made and a painful awareness of the many important questions that remain unanswered. I hope that this book will be a starting point for further progress on methodological problems relating to extrapolation.

Many people deserve thanks for help without which this book would never have come to be. The biggest thanks are due to Sandra Mitchell, who, as my dissertation director, helped me to shape my disorganized ideas into a coherent research project. I am also grateful to Richard Scheines for carefully reading and discussing chapter drafts of that dissertation. There are also a number of people who provided essential assistance in transforming the dissertation into this book. Chief among these is Kristin Schrader-Frechette, whom I met when she came to present a paper at the Michigan State University Philosophy Department. It was largely thanks to her encouragement and support that Oxford University Press agreed to review the manuscript. I'm also grateful to Megan Delehanty for alerting me to the existence of the critiques of animal extrapolation written by Hugh LaFollette and Niall Shanks. Although I disagree with almost all of LaFollette and Shanks's arguments about animal extrapolation, I found them to be the single most useful philosophical source on this topic. It was largely through reading their work that I was able to clearly articulate in simple terms the basic challenges that any adequate account of extrapolation has to surmount. I am also grateful to all those who read and commented on various portions of the manuscript at various stages of completion, including Megan Delehanty, Jason Grossman, Jim Woodward, and Francesco Guala. I would also like to thank the two people who refereed the manuscript for Oxford for their helpful suggestions, and one of those referees especially for suggesting that I change the title. (The original title was *Causality and Heterogeneity*. Try saying that three times fast!) Thanks are also due to audience members for helpful questions asked at presentations at Michigan State University, the University of California at Irvine, the Central Division of the American Philosophical Association, the Center for Philosophy of Science at the University of Pittsburgh, and the Minnesota Center for Philosophy of Science. Finally, I would like to thank Peter Ohlin for prompt and wise editorial advice.

In addition to people, there are some institutions that deserve thanks. I am thankful for the semester of contractual research leave provided by the Department of Philosophy at Michigan State University, and for the Intramural Research Grant Program at Michigan State that enabled me to extend that semester of leave into a full academic year. Without that precious time for reading, thinking, and writing, this manuscript would still be unfinished. I spent that year of research leave as a fellow at the Center for Philosophy of Science at the University of Pittsburgh, which is as about as good an environment for writing a book about philosophy of science as a person could imagine.

Although the bulk of this book is published here for the first time, some chapters contain material from articles that I have previously published. Section 4.4.2 is mostly an abridged version of "Homogeneity, Selection, and the Faithfulness Condition," *Minds and Machines* 16: 303–17. © 2006 by Springer Science+Business Media B.V. All rights reserved. The HIV replication diagram in Chapter 4 (Figure 4.1) and a good deal of Chapter 7 were originally published in "Can a Reductionist Be a Pluralist?" *Biology and Philosophy* 19: 55–73. © 2004 by Kluwer Academic Publishers. All rights reserved. Most of sections 8.2.1 and 8.3.3 originally appeared in "Methodological Individualism, Explanation, and Invariance," *Philosophy of the Social Sciences* 36: 440–63. © 2006 by Sage Publications. All rights reserved. Finally, Chapter 9 is a revised and expanded version of "Social Mechanisms and Causal Inference," *Philosophy of the Social Sciences* 34: 55–78. © 2004 by Sage Publications. All rights reserved.

Contents